Albert F. (Albert Franklin) Blaisdell

## Our Bodies and How We Live

An Elementary Text-Book of Physiology and Hygiene for Use in Schools with Special

Reference to the Effects of Alcoholic Drinks, Tobacco and Other Narcotics on the

Bodily Life

Albert F. (Albert Franklin) Blaisdell

**Our Bodies and How We Live**

*An Elementary Text-Book of Physiology and Hygiene for Use in Schools with Special Reference to the Effects of Alcoholic Drinks, Tobacco and Other Narcotics on the Bodily Life*

ISBN/EAN: 9783744670005

Printed in Europe, USA, Canada, Australia, Japan

Cover: Foto ©berggeist007 / pixelio.de

More available books at **www.hansebooks.com**

# OUR BODIES

AND

## HOW WE LIVE

AN ELEMENTARY TEXT-BOOK OF PHYSIOLOGY AND HYGIENE FOR
USE IN SCHOOLS WITH SPECIAL REFERENCE TO THE
EFFECTS OF ALCOHOLIC DRINKS, TOBACCO
AND OTHER NARCOTICS ON
THE BODILY LIFE

BY

ALBERT F. BLAISDELL, M.D.

AUTHOR OF " HOW TO KEEP WELL " "CHILD'S BOOK OF HEALTH," ETC., ETC.

*REVISED EDITION*

BOSTON, U.S.A.
GINN & COMPANY, PUBLISHERS
1896

# PREFACE

---

It has been the aim of the author in making this text-book, for the use of common schools, to present clearly and tersely the most important and interesting facts concerning our bodily life. To keep any machine in good running-order, we must know something of the structure and use of its various parts. So when we come to the study of this wonderful and complex machine, the human body, we must master such facts of its anatomy and physiology as are essential to a proper understanding of the simple laws of health.

The real object of studying physiology in schools is to teach young folks how to keep well and strong, and to avoid evil habits that destroy character as well as health. Hence, special emphasis has been laid upon such points as have a bearing upon our personal health.

It is now generally admitted that actual observation and actual experiment are as necessary in the study of elementary physiology as in any other branch of science. We must handle the skeleton itself if we wish to get a fair idea of its bones. We must examine a drop of blood under the microscope if we would know how blood looks. Hence, as supplementary to the text itself, we have added a series of practical experiments. Most of them are simple, and require only a little painstaking and an inexpensive apparatus to do successfully. These experiments should be actually performed, and not merely read as a part of the text.

This book has been thoroughly revised. New sections and new chapters have been added, while others have been rewritten and rearranged. The new chapter on Physical Exercise, supplemented by a series of gymnastic exercises, will prove an attractive feature to those who realize the importance of physical culture to our young folks.

The author has aimed to make everything plain by using a simple and homely style of writing.

In the preparation of this revised and enlarged edition, the author and publishers are under deep obligations to Mrs. Mary H. Hunt, the Superintendent of the Department of Scientific Instruction of the National Woman's Christian Temperance Union, who has carefully revised the book.

ALBERT F. BLAISDELL.

JUNE, 1892.

# CONTENTS

---

# OUR BODIES

---

## CHAPTER I

### INTRODUCTION

**1. Animal Life.** — We need not be told that we have bodies made of flesh and bone, whose substance we see and handle every day of our life. Even a child knows that certain parts of his body, as the walls of his chest, and his heart that he feels beating within it, are always moving of themselves.

We can move of our own free will from one place to another. The wind may rustle the leaves, and a breeze may sway a strong oak ; but the leaves and the oak have no power to move of themselves. We need not wait, like the trees of the forest, for the wind to blow us to and fro, or, like the pebbles by the roadside, for some one to stir us. Like the horse, dog, bird, or any other animal that has life, we can move from place to place.

Every living animal, from ourselves down to the tiniest creature that lives its brief life in a single day, and that cannot be seen without the aid of the microscope, has the power to move of itself.

Again, every child must see for himself that he is warm. Even in the coldest day of midwinter, let the stones and trees be as cold as the winter wind, our bodies, except perhaps the tips of the fingers and toes, are always warm. The horse, dog, and even the birds and the bees, are warm. All animals, in fact, are more or less warm as long as they are alive. ,

2. The Body compared to a Locomotive. — Our bodies are in some respects very much like the locomotive. The bones and muscles answer to the machinery of the engine, and the motive-power is produced by the food we eat. We put fuel into the furnace. The water in the boiler is heated and expands into steam. Then the piston begins to work to and fro ; this moves the wheels, joints, and levers : and so the whole engine is set going by the fuel which is put into the furnace.

Now, just the same thing occurs in our bodies. We take food, and that food passes into the stomach. By reason of that food, we are kept warm, muscular force is developed, and the levers and joints within us are set working, as we see in the locomotive.

There is, however, an important difference between the two. The locomotive is all the time wearing itself out. It must be stopped, taken to pieces, and repaired by the machinist. So our bodies are all the time wearing out, but they are constantly repairing themselves. We take food, not only to warm us, but also for the building-up and repairing of our bodies. Remember, then, that the body far surpasses the engine in the perfection of its mechanism, inasmuch as it is self-repairing

3. The Body and its Nervous System. — If we exert our will, or "make up our mind" as we call it, to cross the room, to write with a pen, to pick a flower, to

eat an apple, or to repeat some familiar maxim, we can do it. The mind, or brain, wills to do this or that thing. It sends out its order by the tiny, white, thread-like cords called nerves; and the muscles are put in motion to do whatever the brain has willed them to do.

Every child must have noticed, that when he pricks his fingers with a needle, touches a hot stove, or has eaten some unwholesome food, like an unripe apple, pain is produced. These same silvery-white cords, like telegraph-wires, carry special despatches to the central station; that is, the brain: and the feeling of pain is there made known.

This is the way every part of our bodies is watched over and protected. If it were not so, we should be continually hurting ourselves. This mind, which feels and thinks, but which we cannot see, forms the essential part of our being. It is the power to feel, to know, to think, to reason, and to will, that makes us what we are. Our bodies are ever busy. We eat, drink, and sleep. We move about, and are warm. We feel, hear, see, talk, breathe, and think. We do not wonder about it, simply because it is so common.

When we do stop to think about it, how wonderful it is simply to be alive! What study could be more important, interesting, and even fascinating, than that which has to do with the working of our own bodies! We gaze with wonder and admiration at the marvellous work of some intricate machine made by man. Well we may, for it seems endowed with life. But, in the human body, we find, the more we study and think about it, not simply a most perfect and delicate machine, but one endowed with life, — a mind, — a soul.

Indeed, the more we study, and the longer we live, and reflect upon it, the more we shall realize the great fact that the All-wise Creator, in his goodness and wisdom, has made for us bodies, which, in the words of the Psalmist, are "fearfully and wonderfully made."

4. **Anatomy, Physiology, and Hygiene.** — Before we can tell how plants and animals *live*, we must know what animals and plants *are*. A watchmaker would be unable to describe the working of a watch or clock unless he first made himself acquainted with the various parts of which watches and clocks are made. So it is with the study of human beings. We must know the structure of their bodies before we can understand the manner in which these bodies act and move, or, in one word, *live*.

The science which tells us about the structure, form, and position of the different parts of the body, is called **anatomy**.[1] It tells us *what* they are, *where* they are, and *how* they look.

The science which explains the uses or functions of the different parts of the body is called **physiology**, meaning "a story about nature."

Now, after we have learned something about the structure and uses of different parts of the body, we should learn how to take care of these parts, and how to keep them in health. We do this by the study of **hygiene**,[2] or the science which tells us about health.

---

[1] The word "anatomy" comes from the Greek, meaning "a cutting through," or dissection ; that is, the act of cutting an animal in pieces for the purpose of study.

[2] The word "hygiene" is derived from the name of the Greek goddess Hygeia, who is said to have watched over the health of the people.

Take the stomach for an illustration. If we learn what it is, where it is, how it looks, its shape, size, and general appearance, its coats, etc., this would be its anatomy.

If we learn for what special purpose it is made, just what these different parts do, and how they do it, this would be its physiology.

Finally, if we learn what might interfere with the proper working of each part, and how it is kept in good order, — what will injure its health, and what will do it good, — this would be its hygiene.

5. Some Technical Words Explained. — A child soon learns by experience many things about such parts of his body as the eye, the ear, the nose, the hands and feet, the tongue, and so on. Now, you must already know that there are other distinct parts inside of our bodies, although we cannot see them. You have doubtless heard of some of these parts, such as the heart, the brain, the stomach, the liver, and the lungs. It is usual to speak of these as the organs of the body.

An organ is a special part of the body which performs a special work. Thus the eye is the organ of sight, the nose of smell, the ear of hearing, the stomach of digestion, the lungs of breathing, and so on.

A number of organs, similar in structure, of different size and shape, extending throughout the body, is called a system. Thus we speak of the nervous system, the arterial system, etc.

Several organs, different in structure, but so arranged as to work together for some special end are called an apparatus. Thus the stomach and liver make up a part of the digestive apparatus.

The special work which an organ has to do is said to be the **function** or use of that organ: thus it is the function of the eye to see, and of the liver to secrete bile.

We speak of the body as being an organized structure because it consists of a number of organs, each performing some function necessary to the whole. One important fact must be borne in mind. Although each of these organs is placed in a distinct part of the body, and has its own special work to do, yet no one exists and works for itself alone. The special work which each performs is necessary for the general well-being of the body as a whole, and cannot be carried on by another organ.

Some of the organs are more intimately connected with the life of the body than others. We often speak of them as the **Vital Organs.** If any one of these be diseased or injured, so that it cannot carry on its proper work, the rest of the organs sooner or later suffer too, cease to perform their functions, and death ensues.

FIG. 1. — Portion of a lining tissue showing its bundles and filaments of fibrous tissue, crossing in every direction. The rounded bodies represent a few fat-cells. Highly magnified.

The organs and other parts of the body are composed of a variety of substances or materials. These we shall call the **tissues** of the body. They may be compared to the timber, stone, bricks, mortar, iron, lead, glass, and other materials, which, properly arranged, make up a dwelling-house. The principal tissues which make up the body are *bone*, or *osseous tissue;* *muscle*, or *muscular tissue; cartilage*, or *gristle; fat* and *nerve tissue.*

6. **Minute Structure of the Body.** — When we examine any part of the body with the microscope, we find it made up mainly of **fibres, fluids, and cells.** The original element out of which every other element of the body is formed is the **cell.** The **fibres** serve to give firmness to the tissues or organs and to bind the most minute parts of the body together. Fibres are properly whitish or yellowish threads, slender as one can imagine.

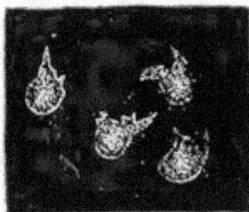

FIG. 2. — With blood-cells in rapid motion, as seen under the microscope.

Some are elastic ; others are not, but very tough and strong.

The *cells* are so small that we must use a powerful microscope to see them at all. In a general way we may compare them to the tiniest bags filled with a semi-fluid mass, in which another still smaller bag floats, and is called the *nucleus* of the larger one. Cells are of all shapes and sizes, round, flat, thick, and long. Thus, we shall learn in a succeeding chapter of the rounded cells, or corpuscles, which float in the blood, and of the flattened cells which can be scraped from the outer or scarf skin.

The life of a cell is one of ceaseless change. It is ever changing its form. If we watch the cells under a microscope, we find them dividing themselves into parts with the greatest rapidity. In their never-ending changes, they are doing a remarkable work in our bodies.

Some change their forms, and become muscle : others become fluids, which help digest our food.  The liver-

cells manufacture or secrete the bile, and the bone-cells help make the bones.  Mil-

FIG. 3. — Showing how a cell divides into two new cells, each with its nucleus, as seen under the microscope.

lions of blood-cells do their work and perish every day, while the brain-cells act in some mysterious way to help us think.  In short, our very life exists in the cell.

7.  **Chemical Elements found in the Body.** — If we put aside the microscope, and call upon the chemist to help us, he will tell us that there are about seventy different elements known to chemistry.  Of these, only sixteen exist in our bodies.  The most important are carbon, hydrogen, nitrogen, oxygen, sulphur, iron, phosphorus, soda, potash, lime, and magnesium.  Thus, there is iron in the blood, lime in the bones, soda in the bile, and potash in the muscles.

Very few of these elements exist as such in the body.  They are mainly found as chemical compounds.  Thus, water, a combination of oxygen and hydrogen, forms about two-thirds of the weight of the body ; and the average human body contains about six ounces of common salt, which is a form of soda.

8.  **General Build of the Body.** — If we look at the human body, we cannot help noticing that it is made up of a middle, nearly round portion, which is called the **trunk.**   On the top of this is placed a kind of round ball, called the **head;** while two pairs of branches, called the **extremities,** are attached to the upper and lower corners of the trunk.

Again, we notice that the whole body has an outer covering, which is called the *skin*. Underneath the skin lie soft masses of red flesh, called the *muscles*. These muscles are mostly fastened to the hard parts, or *bones*. These bones make up the framework of our bodies. There are about two hundred separate bones in the body, which they serve to support. Taken together, they form the **skeleton.**

The limbs are solid ; but the head holds an important organ called the *brain*, which is the centre of the nervous system. A delicate, silvery-white cord, called the *spinal cord*, runs from the brain down the middle of the backbone, and sends its thread-like branches, called *nerves*, all over the body. It is through the nervous system that we are able to think or feel, or, in fact, to know anything.

The trunk is divided into an upper and a lower room by means of a movable partition called the *diaphragm*.

FIG. 4. — Side view, showing the Diaphragm separating the Chest and Abdomen.

The upper room is called the **chest,** and holds the heart and the lungs.

The lower room is called the **abdomen,** and holds the stomach, liver, intestines, and other vital organs.

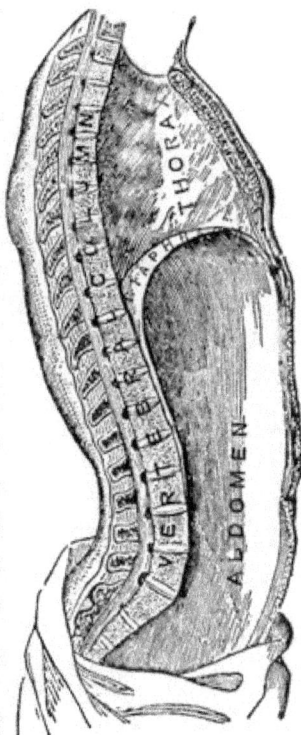

9. **General Plan of Study.** — We shall now proceed
to study some of the most simple facts about the human
body, together with the laws of Hygiene that must be
obeyed to insure its health.

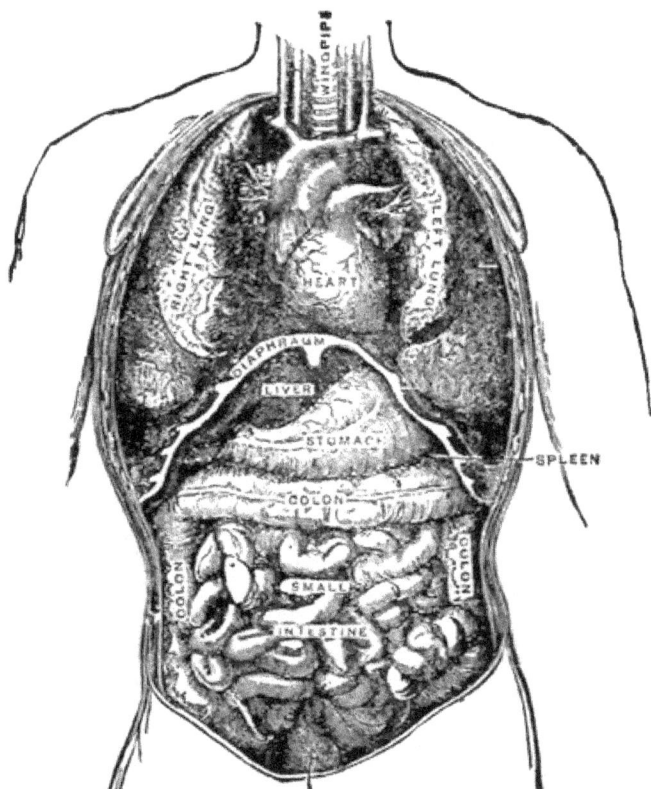

FIG. 5. — Contents of the Chest and Abdomen.

Special emphasis will be laid upon the nature and the
effects of alcoholic drinks, tobacco, and other narcotics.

Reference will be made to the value and importance
of physical culture.

For convenience the whole subject will be considered
under the following general topics : —

## GENERAL TOPICS.

1. **The Skeleton, or bony framework.**
2. **Muscles, or the fleshy parts.**
3. **Physical Exercise in Schools.**
4. **Food and Drink.**
5. **Origin and Nature of Fermented Drinks.**
6. **The Digestion of Food.**
7. **The Blood and its Circulation.**
8. **The Respiration, or Breathing.**
9. **The Skin, or Outer Covering.**
10. **The Nervous System and the Brain.**
11. **The Special Senses.**
12. **Excretion, or getting rid of waste matter.**
13. **The Throat and the Voice.**

1. **The skeleton,** or bony framework, consists of about two hundred bones of various shapes and sizes. It forms the main support of our bodies, and serves to protect the vital organs within.

2. **The muscles** are the fleshy or lean parts of the body ; and they, together with the fat, give to the body its general form and proportion.

3. **Physical exercise in Schools.** Its practical importance and value in promoting the health of young people.

4. **Food and drink** are the materials necessary to keep up the warmth and growth of the body, and enable it to make its movements.

5. **Origin and Nature of Fermented Drinks.** Alcoholic liquors described, and their injurious effects upon the bodily life fully discussed.

**6. Digestion** is the process of changing the food into a condition suitable for absorption into the blood.

**7. The blood** is the vital fluid of the body. The course of the blood through the body is called its circulation.

**8. Respiration** is the act of breathing. It is carried on mainly by the lungs.

**9. The skin,** or the outer covering of the body.

**10. The nervous system** enables us to move, see, feel, taste, and so on. The brain and the spinal cord are its central organs.

**11. The special senses** enable us to touch, taste, smell, hear, and see. We often speak of them as the five senses, or the five gateways of knowledge.

**12. Excretion** is the process by which the body gets rid of its waste matters, mainly through the lungs, skin, and kidneys.

**13.** The **throat** and **voice** described.

## CHAPTER II

### THE BONY FRAMEWORK

10. **The Skeleton.** — Most animals must have some kind of a framework or support to give the body form or shape. This framework in many animals is a support made up of a number of firm and hard substances called bones.

The bones may be on the outside, like a coat of mail, as seen in the lobster and turtle; or there may be a framework of those inside of the body, moved by muscles outside, as seen in the codfish and shad. In fact, all animals that belong to the great backbone family have an inside framework.

This bony support is called the **skeleton,** meaning a "dried up body." It is to the body what the ribs are to a vessel, or what the frame is to a house. Every one is familiar with the picture of the human skeleton. It shows us how the dead bones of the human body look when taken out, cleaned, and held by wires in their proper places. The human skeleton is made up of about two hundred separate bones,[1] of various sizes and shapes.

[1] To be more exact, 28 bones are counted in the head, including the three bones in each ear, 54 in the trunk, and 124 in the limbs. This number depends upon the time and how the bones are counted.

The 32 teeth are not usually reckoned as bones. Several bones, which in youth are made up of separate pieces, unite in old age, and make one bone. The breast-bone, for instance, in the child, consists of eight pieces; but in adult life it is made up of only three.

The bones of the body thus arranged give firmness, strength, and protection to the soft tissues and vital organs, and form, as it were, the foundation upon which our bodies are securely built.

11. **How Bone is made up.** — Bone is a hard and strong substance, made up of animal matter united with certain mineral earths, as lime and potash. The earthy part of bone makes up about two-thirds of its weight, and the animal portion the other third.

Put a slender chicken bone in a mixture of two ounces of muriatic acid and one pint of water, and let it soak for two or three days, and the earthy part will be dissolved, leaving only the animal matter. The form of the bone is kept; but it can be easily bent, cut, torn, or even tied into a knot.

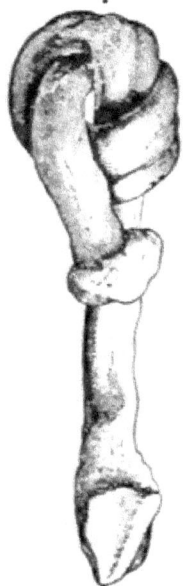

Again, put a soup-bone into a hot fire, and let it burn for two or three hours. Take it out carefully, and, while the shape is still kept, the bone is brittle, and will readily crumble when pinched between the fingers. The animal part has been burned out, and only the white bone-earth is left.

The lime gives hardness and firmness to the bones; while the animal substance makes them elastic, tough, and flexible. The proportion varies with age. In childhood the bones have more animal matter than earthy: hence a child's bones do not break easily, and, when broken, soon

FIG. 6. — The outer bone of the leg, tied into a knot, after the hard mineral matter has been dissolved out by acid.

knit together. In old age, there is more lime in the bones than animal matter: hence they are brittle, and easily broken. They unite slowly, and sometimes not at all.

12. **General Structure of the Bones.** — If we take a long bone, like that from a sheep's leg, or even a part of a beef shin-bone, and saw it lengthwise, we see that

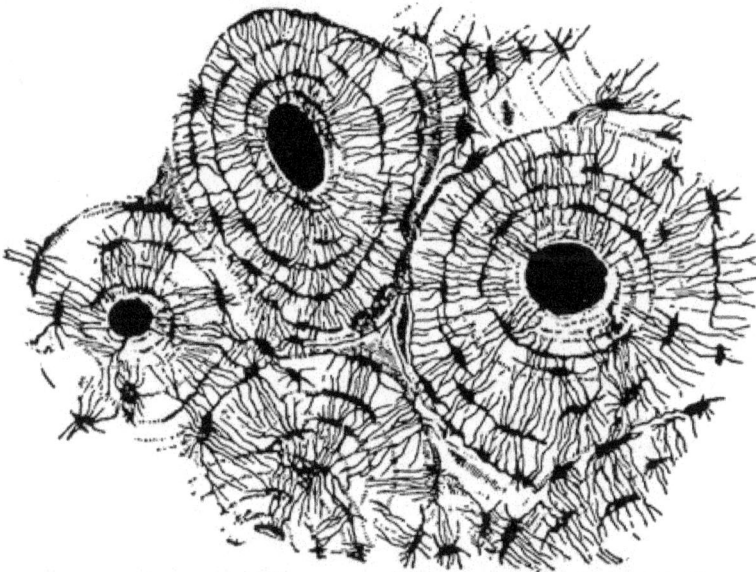

FIG. 7. — A thin slice of bone, cut crosswise, as seen under the microscope.

the ends are soft and spongy, while the shaft is hard and compact. It is hollow, and the central cavity runs almost the whole length of the bone. It is filled, in life, with a soft substance called marrow.

If the bones were solid, they would be much too heavy for ordinary use. A bone may be hard as a rock on the outside, on account of its thin, dense layer of compact bony tissue, and yet be light because of its cavity and the trellis-work of loose, spongy texture at the ends.

Bones are of many different shapes, according to the uses to which they are put. Some are long, with hollow shafts, as the bones of the arm and leg; others are short, to give strength, as the bones of the fingers and toes; some are flat, for protection, and to cover cavities, like the bones of the skull and shoulder-blades; while others are of various odd shapes, and hence called irregular, as the bones of the backbone and ankle.

13. **Minute Structure of Bones.** — When bones are alive or fresh, they have a pinkish-white color, caused by the blood which flows through every part of them. Take a long, slender bone, like a rib or a sheep's leg, and we find that it is elastic, and will bend.

Now, if we scrape down on the bone with a sharp knife, we find that it is wrapped round with a tough but very thin covering, called the *periosteum*, meaning

**Power of Bone to resist Decay.** — "The power of bone to resist decay is remarkable. Fossil bones deposited in the ground long before the appearance of man upon the earth have been found to exhibit a considerable portion of cartilage. The jaw of the Cambridge Mastodon contained over forty per cent of animal matter — enough to make a good glue — and others about the same. From this we see that a nutritious soup might be made from the bones of animals that lived before the creation of man. The teeth resemble bone in their structure, but resist decay longer; they are brought up by deep-sea dredging, when all other parts of the animal have wasted away. The bones differ at different ages, and under different social conditions. In the disease called 'rickets,' quite common among the ill-fed children of the poor in Europe, but somewhat rare in America, there is an inadequate deposit of the mineral substance, rendering the bones so flexible that they may be bent almost like wax. Exercise is as necessary to the strength of bone as to the strength of muscle; if a limb be disused, from paralysis or long sickness, the bones lose in weight and strength as well as the soft parts. Bone is said to be twice as strong as oak, and, to crush a cubic inch of it, a pressure equal to 5,000 pounds is requisite."

"about a bone." All bones are covered with perios-teum except where they go to form a joint. They would die if this covering were for any cause removed.

Bones have a great number of blood-vessels, which, becoming very small and hair-like, pass through tiny holes, or tubes, in the bones, and bring materials for their nourishment and growth. The tiny tubes called "canals," through which the blood-vessels of the bones travel, can be seen only with a microscope. Round these tubes the bony matter is arranged in layers, and between these layers are small cavities. These cavities are joined to each other, and to the canal which they surround, by other extremely fine tubes.

As seen under the microscope, the appearance is not unlike a small animal with a great many legs.

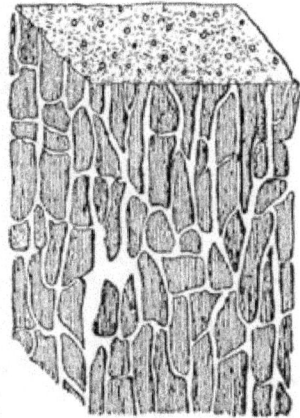

FIG. 8. — Cross Section of Bone, highly magnified; the Canals are left white.

Thus the bones are tunnelled, like honey-comb, with passages through which the blood — "the river of life" — flows in every direction. (Fig. 7.)

### THE HEAD.

14. **The Head.** — The skeleton, or bony framework of the "house we live in," consists of the bones of the **head,** the **trunk,** and the **extremities.**

The bones of the **head** include those of the *cranium* and those of the *face.* Together they form the strong box of bone commonly called the skull.

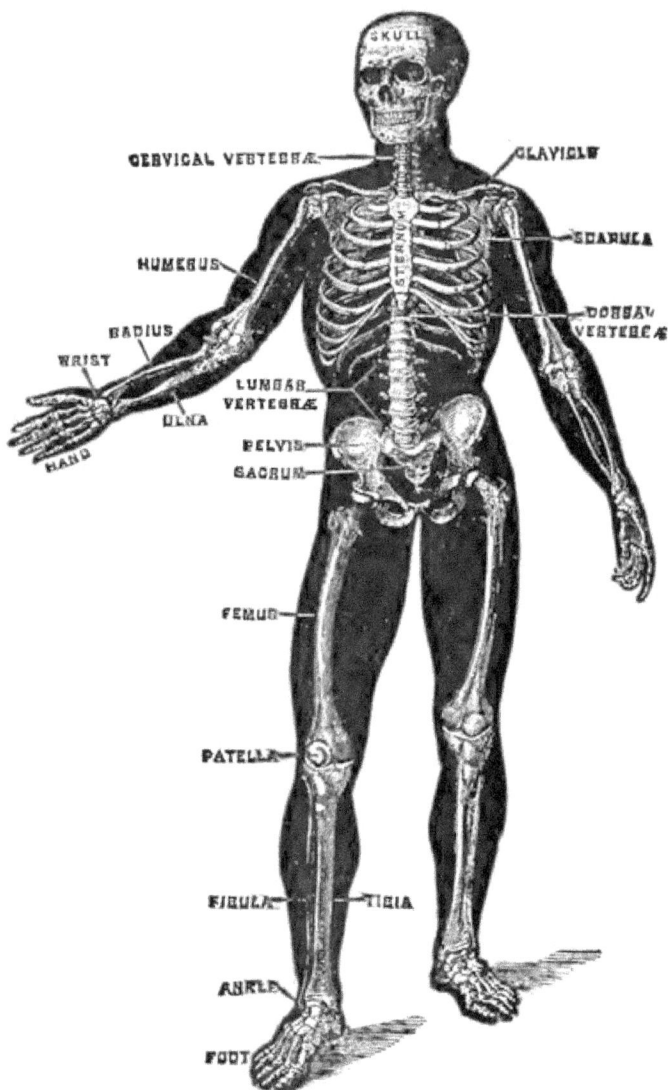

FIG. 9. — Skeleton of Man.

The general shape of the head is that of an arch. The arch is the strongest shape in which the skull could be made, just as the arched bridge is the strongest-shaped bridge which can be made to bear the heavy loads that have to pass over it.

15. **The Cranium.** — The Cranium, or brain-case, is a kind of oval, bony shell, which holds and protects the

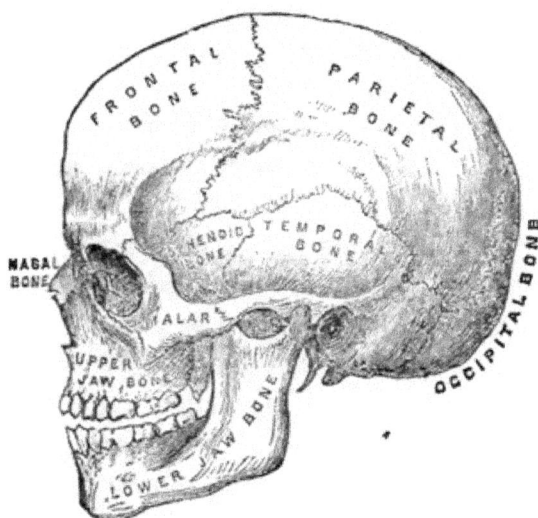

FIG. 10. — The Skull.

brain. It is made up of eight bones closely locked together by seams or sutures, somewhat like the dovetailing used by carpenters. Let us arrange and describe these eight bones thus : —

One *Frontal* (forehead).          One *Occipital* (back of head).
Two *Parietal* (side of head).     One *Sphenoid* (wedge-shaped).
Two *Temporal* (temples).          One *Ethmoid* (sieve-like).

The frontal bone forms the forehead. It is this bone which gives a beauty of form and a dignity of person seen in no other animal in creation.

Two bones make the sides, or walls, of the head. They are called the **parietal** bones, from a Latin word meaning a wall. They join the two **temporal** bones, which lie round each ear, and form the "temples."

The **occipital** bone forms the lower and back part of the skull. This broad, flat bone rests on the topmost bone of the backbone, and is pierced by a large hole through which a long cord of whitish marrow, called the spinal cord, passes from the brain-case into the spinal canal, and runs down the whole length of the backbone.

The **sphenoid,** or wedge bone, is wedged in between the bones of the cranium and those of the face, and serves to lock together fourteen of these bones.

The **ethmoid,** or sieve-like bone, so called because it is full of holes, like a sieve, lies between the eye-cavities just at the root of the nose. The nerves of smell pass through the holes in this bone into the cavities of the nose.

16. **The Face.** — All the bones of the **face,** except the lower jaw-bone, are firmly fixed to each other, and to the bones of the cranium. By their union, these bones form five cavities ; namely, the two large cup-like cavities called the orbits, the two nostrils, and the mouth. The orbits are hollow, bony sockets in which the eyes are placed. In the back part of each cavity is a crevice through which the nerve of sight (optic nerve) passes from the brain to the eye.

The face contains fourteen bones ; viz., —

Two *Malar* (or cheek) *bones.*
Two *Nasal* (or nose) *bones.*
Two *upper Jaw-bones* (upper maxillary).
One *Lower Jaw-bone* (lower maxillary).

Two *Palate bones.*
Two *Lachrymal bones.*
One *Vomer* (ploughshare) *bone.*
Two *Turbinated* (spongy) *bones.*

Under the orbits, and overlapping the jaw-bones, we find the two **malar,** or cheek bones, which in some races, as the American Indians, are very prominent. The two bones that form the upper part, or bridge, of the nose, are called **nasal** bones. The lower part of the nose is gristle, and not bone.

Next we come to the two **upper jaw-bones,** containing, as we all know, some of our teeth. The rest of the teeth are fixed in the **lower jaw-bone,** which moves by means of a hinge-joint, so as to allow the opening and shutting of the mouth.

The remaining bones of the face are small. Two bones, forming the back part of the roof of the mouth, are called the **palate** bones. Two little bones, partly forming the inside walls of the eye-cavities, are called the **lachrymal** bones, from a Latin word meaning a tear. There is, through each of them, a canal which carries the tears from the eyes to the nose.

The **vomer,** or ploughshare bone (so called from its resemblance to the share of the farmer's plough), is situated between the nostrils. Two delicate little bones within the nose-cavity, shaped like a scroll or horn, are called **turbinated** bones, from a Latin word meaning a whirl.

**17. How the Bones of the Head are joined together.** — The bones of the head are joined together in a peculiar way. The edges of the bones of the cranium and face are shaped somewhat like the teeth of a saw.

In adults, these edges fit into each other, and grow together, resembling the dovetailed joints in a cabinet-maker's vork : hence they are called *sutures*, from a Latin word which means a sewing or a seam.

In infancy all of the bones of the skull do not meet, and the throbbing of the brain at the top of the head is easily seen. These openings are called *fontanelles*, "little fountains." The bones of the skull are not wholly welded together till the child reaches adult life. When the bones are knitted together, these sutures add greatly to the strength and resistance of the brain-case.

### THE TRUNK.

**18. The Trunk.** — The trunk is that part of the body which supports the head, and to which the arms and legs are attached. It has two important rooms, or cavities.

The upper one, called the **thorax,** or chest, consists of a bony framework formed by the breast-bone, ribs, and backbone. It contains the *lungs*, with which we breathe ; and the *heart*, which pumps the blood through the body.

The lower part, or **abdomen,** holds the *stomach, liver, bowels, kidneys*, and other important organs.

Between the chest and abdomen is stretched a broad muscle, which moves up and down as we breathe.

This is called the "midriff" by the butchers, or **diaphragm,** from a word meaning a fence or a partition.

The principal bones of the trunk consist of those of the **spine,** the **ribs,** and the **hips.**

There are fifty-four bones in the **trunk,** and they are thus arranged :—

*I.   The Spine contains 26 separate bones.*

- *7 Cervical* (or neck) *Vertebræ.*
- *12 Dorsal* (or back) *Vertebræ.*
- *5 Lumbar* (loins) *Vertebræ.*
- *I Sacrum,* or sacred bone.
- *I Coccyx,* or cuckoo-bone.

*II.   The Ribs,* 24 bones .

- *7 True Ribs.*
- *3 False Ribs.*
- *2 Floating Ribs.*

*III.   I Sternum,* or breast-bone.
*IV.   The 2 Hip-bones.*
*V.   The Hyoid bone.*

19.   **The Spine.** — The **spine,** or backbone, serves as a support for the whole body.   It is made up of a number of separate bones called *vertebræ*, between which are placed pads, or cushions, of gristle.

These cushions are elastic[1] or springy, yet thick and strong.   They serve to break the force of any shock or injury which the spine may receive, just as the springs of a carriage lessen the jolting which would be felt without them.   These soft pads also prevent any grating or friction of one bone on another.   They are also yielding ; that is, are readily pressed together, else the· backbone could not bend.[2]

[1] The elasticity of these plates of cartilage is so great, that we are actually about half an inch shorter when we go to bed than we were when we got up in the morning, by reason of their flattening out under the weight of the erect position.

[2] These bones, again, are prevented from yielding too much by strong tough straps, or ligaments, which stretch between them. If the separate bones were allowed too much freedom of movement, the spinal cord inside the canal would be crushed or twisted, and paralysis or death would at once follow.

The spine is really a pile of twenty-four separate
bones, resting on and above a strong,
three-sided bone called the **sacrum**,
or sacred bone, which is wedged in
between the hip-bones. The sacrum,
although reckoned as one bone,
actually consists of five distinct vertebræ, which grow together, however,
and form a single bone in adult age.

Joined to the lower end of the
sacrum is a little, tapering bone called
the **coccyx**, so named from its resemblance to the beak of a cuckoo.

The whole spine forms a pillar, or
column, of bones, tapering as it goes
up. At the top are seven *cervical* or
neck vertebræ : below them are the
twelve *dorsal* or back vertebræ, from
which spring the ribs. The five lowest bones, called the *lumbar*, or loin
vertebræ, are the thickest and largest.

Each vertebra of the backbone is
pierced with a hole through its centre ;
and the separate bones are so placed
one above the other, that these holes
form a continuous tube, or canal, down
which passes the spinal cord.[1]

FIG. 11. — The Backbone.

[1] Imagine a number of spools placed one on another. The central hole through
each would be exactly over the other, and there would be one long tube, or channel,
through the whole string of spools. This is somewhat like the arrangement of
the vertebræ of the backbone.

In this bony canal the spinal marrow lies protected from injury. From each vertebra project spines, or thorns, of bones, to which are fastened muscles, which keep the flexible backbone erect, and lift the head and shoulders. The row of spines along the whole length of the backbone forms a ridge, which we can feel by pressing with the fingers up and down the middle of the back.

The spine is one of the most curious and wonderful things in nature, — so firm, and yet so elastic ; so stiff that it will bear a heavy weight, and yet bending like rubber ; a tapering pile of odd-shaped bones, so admirably planned, and so wonderfully put together, that the delicate brain resting upon it, and the spinal cord hidden within its bony canal, are not often hurt. The most daring athlete rarely breaks the bones of his spine, or puts them out of place.

**20. The Ribs.** — The ribs are the slender, but strong, bony hoops which help make the framework of the chest. They pass around the chest, and strengthen it, resembling somewhat the hoops of a barrel. There are twenty-four ribs, — twelve on each

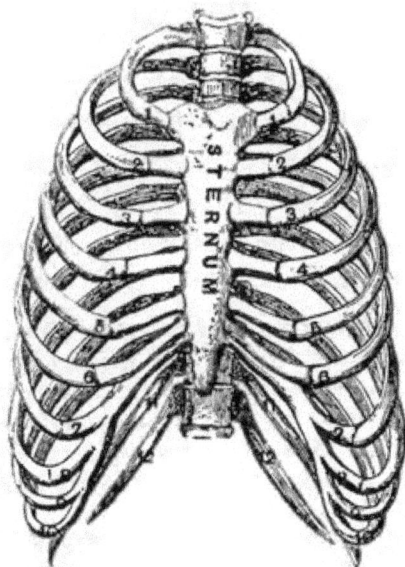

FIG. 12. — The Ribs and Sternum.

side. The ribs are joined to the backbone behind, and in front to a flat, narrow bone, shaped somewhat like an ancient sword or dagger, called the **sternum,** or breast-bone. This bone does not go down far enough to allow of all the ribs being joined directly to it. Only the four-teen upper ribs — that is, seven on each side — are thus fastened : they are on this account called the *true* ribs.

The next six ribs, three on each side, come partly round from the spine, and then each is joined by gristle to the one above it. They are called the *false* ribs. The remaining four ribs, two on each side, spring from the spine, like their neighbors, but are not fastened to any-thing in front, and for this reason are called the *floating* ribs.                                                      .

The spaces between the ribs are filled with strong muscles, one set of which raises the ribs, the other lowers them. This barrel-shaped framework of bones that makes the chest, although strong, is not rigid. It can be raised and lowered, owing to the joints at the back, the cartilages, or gristle, in front, and the muscles between. The space within thus becomes larger, and more room is allowed for the action of the heart and lungs.

**How the Skull and the Spine are joined together.** — The skull and the spine are joined in a peculiar manner. The first, or topmost, of the neck vertebræ is called the *atlas,* so called from the fabled Atlas of the ancients. In olden times, people believed that the earth was supported on the shoulders of a huge giant called Atlas. As, therefore, the head is supported by this bone, it received the name of atlas, after the fabled giant of old.

On the lower part of the skull are two smooth rockers, which fit into two little smooth grooves hollowed out in the atlas. We are thus able to rock the head to and fro. The second vertebra of the neck is called the *axis,* and has a peg, called the odontoid process, or tooth-like peg, which fits into a notch in the atlas. Thus, when we turn the head sideways, the atlas turns around the peg of the axis.

**21. The Hips.** — The lower part of the trunk is formed by two large, irregular bones, very firm and strong, called the hip or haunch bones. They join the

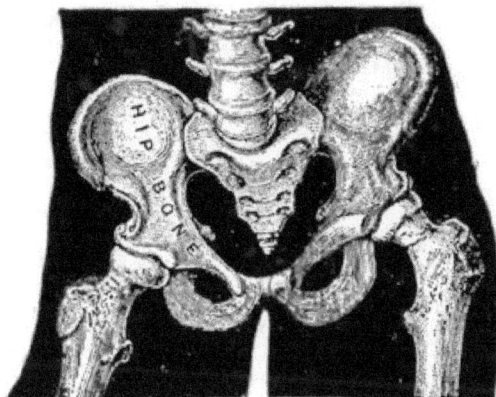

FIG. 13. — The Pelvis.

sacrum behind, and each other in front. The two **hip-bones**, with the sacrum and coccyx, form a kind of bony basin called the *pelvis*, in which many important organs are held. Each hip-bone has a deep, cup-shaped cavity, or socket, into which the rounded head of the thigh-bone fits.

**The Hyoid Bone.** — Under the lower jaw is a little horseshoe-shaped bone called the *hyoid* bone because it is shaped like the Greek letter *v*. The root of the tongue is fastened to its bend, and the larynx is hung from it as from a hook. The hyoid, like the knee-pan, is not connected with any other bone.

### THE UPPER LIMBS.

**22. The upper Limbs.** — Man has two upper and two lower limbs. Each upper limb consists of three parts, the **upper arm,** the **fore-arm,** and the **hand.** Each lower limb also consists of three parts, — the thigh, the lower leg, and the foot.

In speaking of the bones of the upper extremities, however, it is usual to include the two shoulder-blades and the two collar-bones. Each upper limb, therefore, contains the following bones : —

Upper arm . . . . . . $\left\{\begin{array}{l}\textit{Scapula,} \text{ or shoulder-blade.} \\ \textit{Clavicle,} \text{ or collar-bone.} \\ \textit{Humerus.}\end{array}\right.$

Fore-arm . . . . . . . $\left\{\begin{array}{l}\textit{Ulna.} \\ \textit{Radius.}\end{array}\right.$

Hand . . . . . . . . $\left\{\begin{array}{l}\textit{8 Carpal} \text{ (wrist) bones.} \\ \textit{5 Metacarpal} \text{ bones.} \\ \textit{14 Phalanges,} \text{ or finger-bones.}\end{array}\right.$

Making thirty-two bones in all.

**23. The Upper Arm.** — There are two bones in the shoulder, and they serve to fasten the arm to the trunk. These are the **scapula,** or shoulder-blade, and the **clavicle,** or collar-bone. The shoulder-blade is a large, flat, three-sided bone, which is placed on the

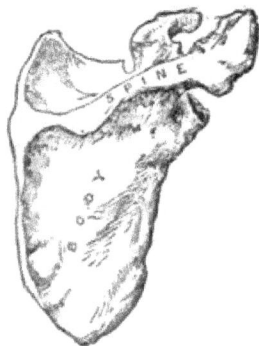

upper and back part of the chest, outside of the ribs. On the outer side it has a saucer-like cavity into which the rounded head of the arm-bone fits.

The collar-bone has a double curve like the Italic letter *f.* It lies across the shoulder, like a kind of rigid bar, over and above the first rib. It

FIG. 14. — Back of Right Shoulder-Blade.

serves, like the keystone of an arch, to keep the shoulders apart, and to hold them firm when the upper limbs are used. **Its** inner end is tied to the breast-bone, and its outer to the shoulder-blade.

The **humerus** is the long, hollow bone of the upper arm. It fits into the socket of the shoulder-blade, and goes to the elbow, where it is joined to the two bones of the fore-arm.

**24. The Fore-arm.** — The fore-arm contains two long, hollow bones, the **ulna** and the **radius.**

The ulna is the larger of these two bones, and is joined to the humerus by a hinge-joint at the elbow. It is prevented from moving too far back by a process, or projection, which makes the sharp point of the elbow.

The radius is the long, slightly curved outer bone of the fore-arm. It is smaller than the ulna. Its upper end is fastened both to the ulna and the humerus. Its lower end is much larger than its upper, and carries the hand. The radius is tied to the ulna in such a manner that it can glide nearly round it. This gives us the power of twisting the fore-arm and the hand.

**25. The Hand.** — The **hand** consists of three parts, the **wrist,** the **palm,** and the **fingers,** containing in all twenty-seven bones.

There are eight **carpal,** or wrist bones, all of which are very closely packed in two rows, and bound together with cords. The wrist is thus as strong as if made of a single bone, while at the same time it is capable of the most delicate movements.

The **metacarpal** bones are the five long bones which form the palm of the hand. They are attached to the carpal bones of the wrist, and to the bones of the fingers.

The **phalanges** of the fingers are the fourteen small
bones which are tied end to end, to form the fingers.
Each finger has three bones, each thumb two. The
bones of the fingers are arranged in three rows, as will
be seen by closing the hand.

The wrist-bones, metacarpal bones, and finger-bones
are all held in place by strong but flexible ligaments.
By this beautiful contrivance, the greatest strength and

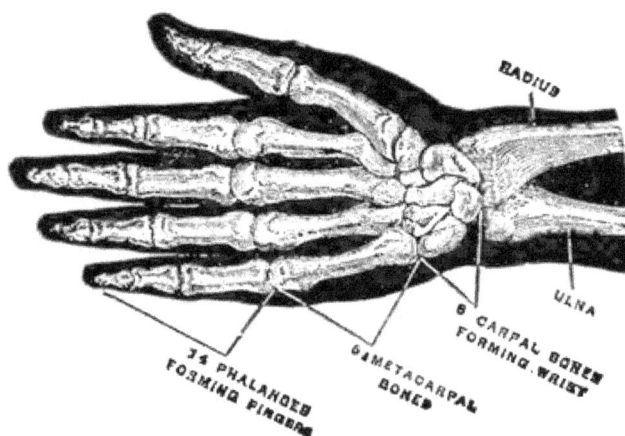

FIG. 15. — Bones of Right Hand.

elasticity are given to the hand, which is thus fitted for
all kinds of work, from grasping heavy hammers to
handling the pen, and threading the finest needle.

### THE LOWER LIMBS.

**26. The Lower Limbs.** — The general structure
and number of the bones of the legs bear a striking
similarity to those of the arms. Thus the leg, like the
arm, consists of three parts, the **thigh,** the **lower leg,**
and the **foot.**

There is only one bone in the thigh, while there are two in the lower leg. We see that this is a similar plan to that of the arm. In the foot, the tarsal or ankle bones correspond to the carpal bones of the wrist; while the metatarsal bones and the phalanges of the foot take the place of the metacarpal bones and the phalanges of the hand.

The bones of the leg are, —

| | |
|---|---|
| Thigh . . . . . . . . . . | Femur. or thigh-bone. |
| Lower leg . . . . . . . . . | { Patella, or knee-pan.<br>Tibia, or shin-bone.<br>Fibula, or splint-bone. |
| Foot . . . . . . . . . . | { 7 Tarsal, or ankle-bones.<br>5 Metatarsal, or instep-bones.<br>14 Phalanges, or toe-bones. |

Making thirty bones in all.

27. **The Thigh.** — The top bone of the leg is the **femur**, or thigh-bone. It is the largest and strongest bone of the body. It has a rounded head, which fits into a cup-like cavity in the hip-bone. At the knee-joint the thigh-bone meets the bones of the lower leg, to which it is fastened by strong cords.

28. **The Lower Leg.** — The lower leg consists, like the fore-arm, of two bones. The larger, a strong, three-sided bone, with a sharp edge in front, is called the **tibia,** and is commonly known as the shin-bone.

The smaller bone, bound at both ends to the tibia, is called the **fibula,** meaning a clasp. It is a long, slender bone on the outside of the leg, and its lower end forms the outer ankle. It is often spoken of as the small bone of the leg.

Covering in part the knee-joint is a flat, three-sided bone, called the **patella,** or knee-pan, which protects the joint, and gives firmness to the leg.

**29. The Foot.** — The **foot,** like the hand, consists of three parts, the bones of which are known as the tarsal bones, the metatarsal, and the phalanges of the foot.

The **tarsal,** or ankle-bones, are seven in number, and form the heel, the ankle, and part of the sole of the foot. These seven irregular bones are tied firmly together by straps, or ligaments, and are strong enough to bear the weight of the body. The large bone,

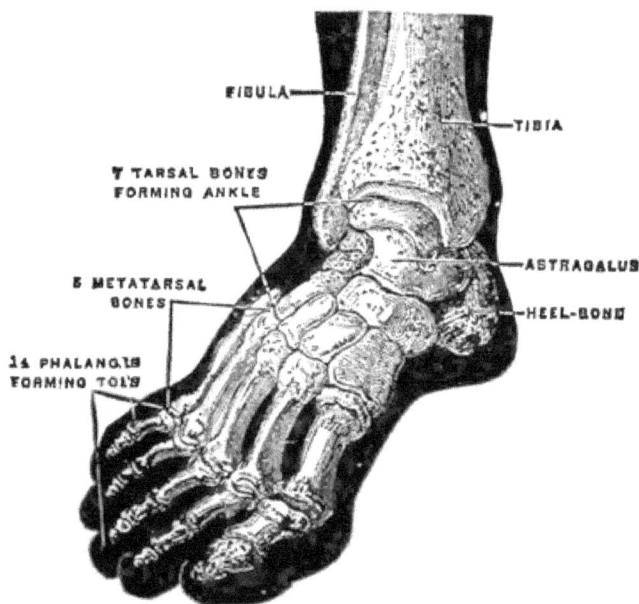

FIBULA

TIBIA

7 TARSAL BONES
FORMING ANKLE

5 METATARSAL
BONES

ASTRAGALUS

HEEL-BONE

14 PHALANGES
FORMING TOES

FIG. 16. — Bones of Foot and Ankle.

which projects backwards, is the heel-bone ; and on the bone over this the shin-bone rests. The heel-bone is connected with the great muscles of the calf of the leg by a very strong cord, or tendon, called the tendon of Achilles.[1]

[1] The warlike deeds of this famous Greek hero were sung by Homer. According to the story, Achilles received his death-wound in the heel, no other part of his body being vulnerable.

The **metatarsal** bones, like the palm of the hand, are five in number, and form the instep of the foot. They are the connecting link between the ankle-bones and the phalanges of the foot.

The bones of the toes, or **phalanges**, number fourteen in all, three in each toe, with the exception of the great toe, which, like the thumb, has only two bones.

30. **Use of the Bones.** — Bones serve many useful purposes. They keep up the general shape of the body. The skeleton, as we have seen, is its framework. It gives strength and support to the soft and fleshy parts. If there were no bones, and the whole body were a mass of flesh only, the legs would give way, and finally be crushed down, under the great weight of the body.

Again, bones protect the soft organs which lie beneath them. The bones of the head protect the soft and delicate brain in a complete box of bone : the ribs

**The Hand and Foot compared.** — The toes and fingers correspond exactly in their number and the character of their bones, the great toe holding the same place in the foot as the thumb does in the hand. In the hand the wrist-bones are small and the finger-bones long and slender. The thumb is made to move in the opposite direction to the fingers, and thus the hand becomes adapted for grasping purposes. Indeed, the hand may be either a most delicate pincer or a powerful vise, while its flexibility and rapidity of motion can nowhere be better seen than in the rapid movements of the skilful musician on the keyboard of the piano or organ.

Now look at the bones of the foot. The ankle-bones have a clumsy look, and all the hollow bones are short, thick, and heavily jointed. It is at once clear that strength, and not flexibility, is the object of these bones. The feet, which are to support the entire weight of the body, are thus better adapted to do their work than they would be if built of slender, long bones like those of the hand.

protect the heart and lungs in a large cage of bone, and so on. Passage-ways and little cavities are hollowed out of solid bone to lodge and shield important organs. Grooves and canals are formed in the hard bone to receive and protect tender organs, delicate nerves, and tiny blood-vessels.

Finally, the surfaces of bones are fitted with grooves, knobs, and sharp edges, to which muscles are tied. We are thus enabled to stand erect, and make with ease and quickness the countless movements of the body. The blood filters through bone as freely as it does through any other living tissue. The bone-structure is ever changing. Old material is got rid of, and new matter takes its place.

**31. Repair of the Bones.** — When a bone is broken, blood trickles out between the injured parts, and afterward gives place to a sticky, watery fluid, which gradually becomes thicker, like sirup or jelly. This slowly hardens into a new bone-structure, and forms a kind of cement to hold together the broken ends. Nature does not spare her healing cement. The excess bulges out around the place of union, over which a bunch is felt, for a lifetime perhaps, under the skin.

In young people, a broken bone will knit together in two or three weeks ; while in grown-up people six weeks or more will be required. In aged persons, a broken bone may cripple them for life. It is then a tedious matter, and sometimes the bones will not unite at all.

After a bone has been once broken, it is fragile for some time ; and great care should be taken lest it be

broken a second time before it firmly unites. When a
bone is broken, the ends tend "to ride" over each
other, because the muscles pull the broken portions
apart ; hence the need
"to set" the bone by
drawing the injured bone
into place, and keeping
it so by splints and ban-
dages, properly applied
by a surgeon.

32. **How Bones are**
**joined together.** — The
place where two bones
join together is called a
**joint.**

Joints vary according
to the kind and amount
of motion. Get a knuckle
of ham, mutton, or a
beef-joint at the market.
Cut into it, open up the
joint, and study its
structure. In all joints,
the essential parts are
the same. The ends of
the bones are shaped
according to the special
needs of each joint.

Fig. 17. — The Right Knee-Joint, showing how
firmly it is bound about by ligaments.

The joint-end of the bones is smooth, moist, and
tipped with a thin layer of gristle, or cartilage. This
smooth and glistening covering is bathed by a sticky

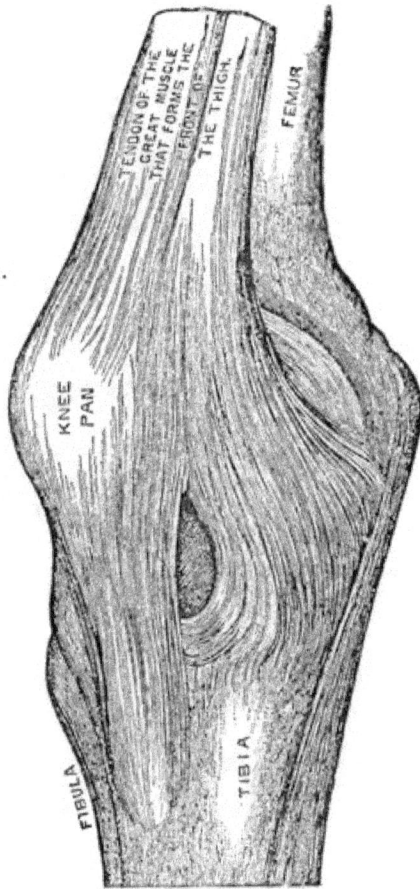

fluid called the synovial fluid, so named because it is like the white of a raw egg. This is the oil often spoken of as "joint-oil," furnished by nature to allow the rubbing-surfaces to move smoothly over one another, and thus prevent too much wear and tear.

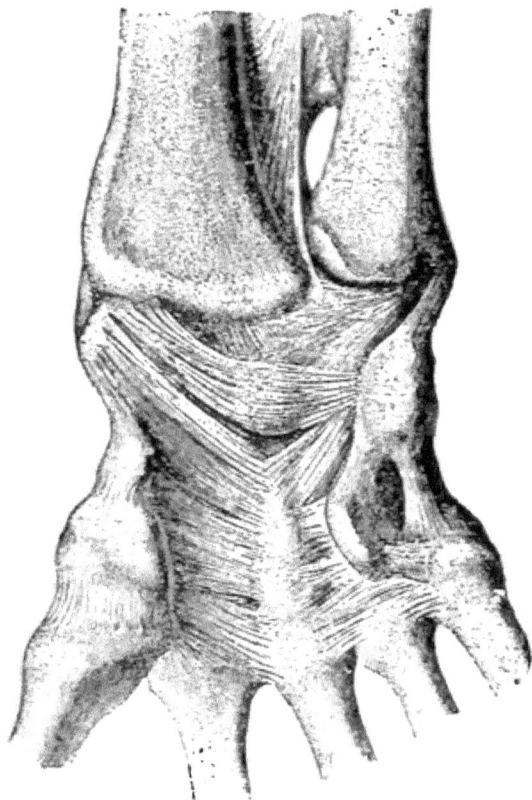

FIG. 18. — Showing how the Wrist-Bones are tied together.

There are two principal kinds of joints, the *fixed* and the *movable*. Thus, the bones of the head, as we have seen, are firmly dovetailed into each other by jagged, saw-like edges, which grow into each other from infancy. The bones of the sacrum are also firmly united. to each other. These are the "fixed" joints.

Movable joints allow the bones to glide on each other with more or less freedom of motion. They differ according to the motion needed. Such a joint as that at the hip is called a "ball-and-socket joint," because the rounded head of the thigh-bone fits into the cup in the hip-bone. The ball-like head of the arm-bone works in the cavity of the shoulder-blade, and makes another ball-and-socket joint. Such joints allow a greater variety of motion than any other kind.

Again, bones are grooved and ridged so that one bone can glide over the other to and fro, like a door on its hinges. This is called a "hinge-joint." Such joints are found at the knee, and between the lower jaw and cranium. The bones of the fingers also move on each other in this way. Sometimes a kind of peg in one bone fits into a notch in another, and forms a "pivot-joint." The two bones (atlas and axis) at the top of the spine are joined in this way.

**33. How Bones are tied to each other. Ligaments.** — The bones are tied together, kept in place, and their movements limited, by tough and strong bands, or straps, called **ligaments,** meaning "to bind." They may be seen in any of the movable joints, — say of the calf, sheep, or chicken, — and have the look of white, silvery cords.

Some of the ligaments are as thin as a piece of paper; while others, as at the side of the knee, or at the shoulder, are very thick. Some cross each other, as in the knee-joint; while others go all round the joint, and completely shut it up in a bag. This prevents the bones from being easily dislocated, or slipped out of place.

It is a difficult matter to carve a turkey or a fowl, because one has to cut through these ligaments before he can cut the limbs apart to serve it out in pieces. There is the same difficulty in separating the two bones in a shoulder or leg of mutton, because they are held firmly together by strong ligaments.

**34. Hints about the Health of Bones.**—The bones of children are flexible, and capable of being bent by long-continued strain; because there is more animal matter than in later years. Therefore great care must be taken with the positions which children take at home, at school, and elsewhere.

FIG. 19.—A School-girl sitting at her Desk in a position often resulting in Curvature of Spine.

Allow a child to walk too early, before the legs are strong enough to bear the weight of the body, and "bow-legs" result.

At school, the desks should not be too low, thus causing a forward stoop; or too high, thereby throwing one shoulder up too much, and giving a twist to the spine. If the seats are too high, the feet have no support, and injury to the thigh may result. If too low, there is undue strain on the shoulder and backbone. Round shoulders and curvature of the spine may result from long-continued positions of this kind.

The feet should rest firmly on the floor, and the edge of the desk should be about one inch higher than the level of the elbows. A line dropped from the edge of the desk should strike the edge of the seat. Three sizes of desks should be used in every schoolroom, and more in ungraded schools. Seats should be regulated according to the size of the pupil, and frequent changes of seats should be made.

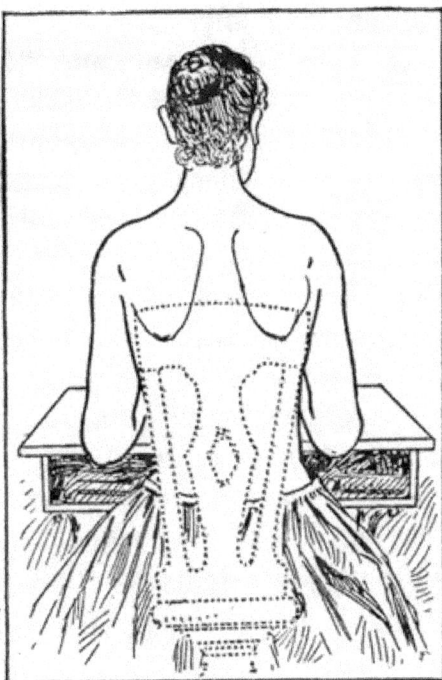

FIG. 20.—A correct Position at the School-desk, with no undue Strain on the Spine.

Young people should not get into the habit of taking hurtful positions, such as sliding down into the seat, sitting on the foot, or on the small of the back. Bending over too much while reading, writing, sewing, or otherwise at work, is apt to cause spinal curvature and round shoulders.

The prevailing fashion of using tight and high-heeled boots and shoes cannot be too strongly condemned as both hurtful and ugly. High heels throw the weight of the body forwards, and force the foot down on to the toes. This will in time not only crowd all shape out of the toes, causing tender feet, corns, bunions, distorted joints, and in-growing nails, but makes the natural gait stiff and ungainly.

**35. Effect of Alcohol and Tobacco on the Bones. —** Since the bones constitute the framework of the body, a person's form depends upon the size and shape of his bones. The bones grow during childhood and youth; whatever growth one loses during that time cannot be afterward made up. It is the testimony of sagacious physicians that alcoholic drinks and tobacco tend to check the growth of the bones.

The smoking of cigarettes is especially hurtful to growing boys, because such a habit tends, besides other harmful things, to dwarf the growth of the bones. A well-developed form is something to be prized. No wise boy or girl will risk attaining it by indulging in filthy or injurious habits while young.

See Note 1, page 355.

**REVIEW ANALYSIS: THE SKELETON—(206 bones).**

| | | |
|---|---|---|
| **THE HEAD (28 bones).** | I. CRANIUM (8 bones) | 1 Frontal (forehead), <br> 2 Parietal (side of head), <br> 2 Temporal (temples), <br> 1 Occipital (back of head), <br> 1 Sphenoid (wedge-shaped), <br> 1 Ethmoid (sieve-like). |
| | II. FACE (14 bones) | 2 Malar (cheek), <br> 2 Nasal (nose), <br> 2 Upper Jaw-bones, <br> 1 Lower Jaw-bone, <br> 2 Palate bones, <br> 2 Lachrymal bones, <br> 1 Vomer (plough-share) bone, <br> 2 Turbinated (spongy) bones. |
| | III. THE EAR (6 bones) | Hammer, <br> Anvil, <br> Stirrup. |
| **THE TRUNK (54 bones).** | I. SPINAL COLUMN (26 bones) | 7 Cervical (neck) Vertebræ, <br> 12 Dorsal (back) Vertebræ, <br> 5 Lumbar (loins) Vertebræ, <br> Sacrum (sacred bone). <br> Coccyx (cuckoo-bone). |
| | II. THE RIBS (24 bones) | 7 True Ribs, <br> 3 False Ribs, <br> 2 Floating Ribs. |
| | III. STERNUM (breast-bone). <br> IV. TWO HIP-BONES. <br> V. HYOID BONE. | |
| **UPPER LIMBS (64 bones).** | UPPER ARM | Scapula (shoulder-blade), <br> Clavicle (collar-bone), <br> Humerus (arm-bone). |
| | FORE-ARM | Ulna (fore-arm), <br> Radius (spoke-bone). |
| | HAND | 8 Carpal (wrist) bones, <br> 5 Metacarpal (palm) bones, <br> 14 Phalanges (fingers). |
| **LOWER LIMBS (60 bones).** | THIGH | Femur (thigh-bone). |
| | LOWER LEG | Patella (knee-pan), <br> Tibia (shin-bone), <br> Fibula (splint-bone). |
| | FOOT | 7 Tarsal (ankle) bones, <br> 5 Metatarsal (instep) bones, <br> 14 Phalanges (toes). |

# CHAPTER III

## THE MUSCLES .

**36. The Muscles.** — The smallest child or the tiniest insect can move its own limbs when it pleases, but the largest vegetable and the sturdiest tree never move except some cause outside of themselves acts upon them.

Movement in man is made by **muscles.** The limbs are moved by muscles. Even the motions of the stomach and the action of the heart are controlled by muscles. Muscles move the skin. In many animals this action is well marked, as when the horse shakes his hide to get rid of biting flies. Muscles move the bones, the fingers, and toes, the mouth, and the eyelids. In brief, all motion in the body is dependent upon them.

**37. The Structure of Muscle.** — Muscle is simply the lean meat, or flesh, of the body. When we eat beefsteak or lean mutton for dinner, we are eating muscle.

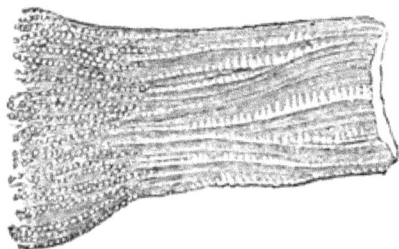

FIG. 21. — A Muscular Bundle teased out to show its Fibres.

Muscles are made up of bundles of fleshy strings called fibres held together by a very thin web of tissue, not unlike the thinnest of tissue paper. Each fibre is made up of a great many smaller fibres, called fibrils, held together in the same way. Each fibril consists of rows of little cells, arranged like a string of beads. This gives the muscles a peculiar striped look under the microscope, which marks the ordinary muscles of the body.

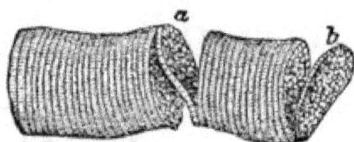

FIG. 22. — A Fibre of Muscle, showing the separate Disks of which it is composed at *a* and *b*.

When corned beef has been "boiled to rags," as we say, the web of connecting tissue is partly dissolved by heat, and the fibres are plainly seen.

**38. How Muscles act.** — Muscles have a peculiar power of their own. This is the power to **contract,** or to shorten themselves.

In order for a muscle to act, it must be stimulated. This stimulus, or that which gives it the power to act, is the nerve-force flowing through the nerves distributed to the muscles.

Contraction is not, however, the natural state of a muscle. After a longer or shorter time, it is tired,

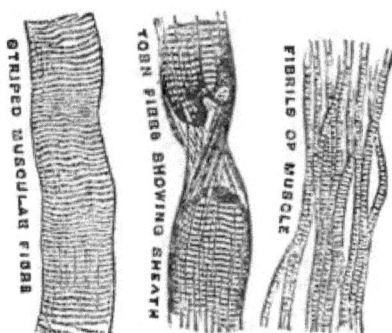

FIG. 23. — Portions of the Muscular Fibre highly magnified.

and begins to relax. Even the heart, the hardest-working muscle of the body, pumping blood night and day all the days of our life, has a short time to rest between its beats.

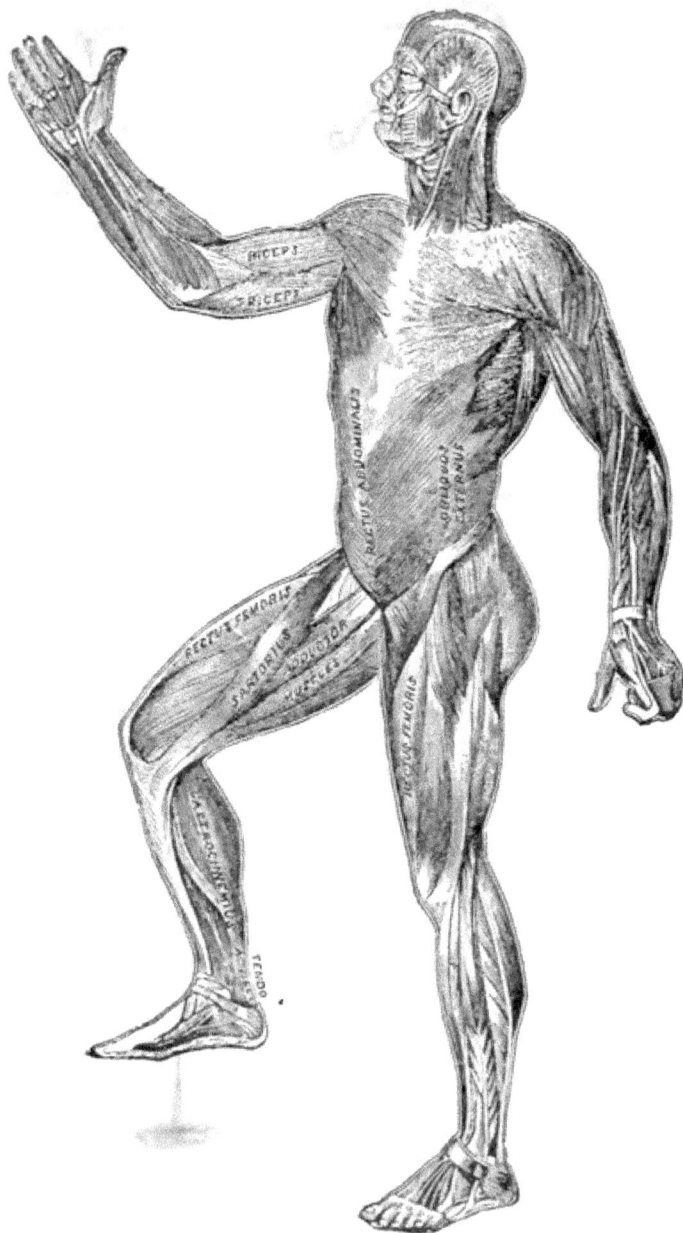

FIG. 24. — Principal Muscles of the Front of the Body.

**39. Kinds of Muscles.** — Muscles are of two kinds, *voluntary* and *involuntary*.

Voluntary muscles are those which can be made to act under the influence of the will. Such muscles make up the bulk of the fleshy parts of the body, and form a covering for the bony skeleton.

The involuntary muscles are those which carry on their work without the use of the will — indeed, we cannot prevent them acting. Among these muscles may be mentioned those of the stomach, the bowels, the lungs, and the heart. When they contract, they force along the contents of these organs. In this way the blood is forced out of the heart, the food out of the stomach, and so on. If these muscles were not entirely independent of the will, they would cease to act the moment we fell asleep, and death would be the result. Hence these muscles continue to work at all times, because the will has no power over them.

**40. Arrangement of Muscles.** — The muscles are arranged, for the most part, so as to oppose each other in action. Thus, when one set produces motion in one direction, there is, opposite to it, another group of muscles, which draws the part moved back into place again.

When a piece of lean meat is brought from the market, it looks like a solid mass of flesh. It really consists of parts of several distinct muscles. Each of these muscles, when separated, will have the following form : —

A middle or thick part. Two tapering ends, one attached to a fixed bone, and called the *origin ;* the other connected with a movable bone, and called the *insertion* of the muscle. These tapering ends are attached, or tied, to the bones by means of sinews, or tendons.

As a muscle acts only by contracting, each movable bone must be supplied with at least two opposing muscles. One of these contracts, and draws up the limb: this is called a *flexor* muscle. The other, on the opposite side, by contracting, stretches out the limb : this is called an *extensor* muscle.

Take, for instance, the two main muscles of the arm. When we bend the arm, we notice that the muscle in front of the upper arm swells up, and becomes hard. This is the biceps, or two-headed muscle, which has contracted in obedience to the will, and by its contraction has drawn up the fore-arm. When we wish to straighten the arm, we do not merely relax this muscle ; for, if we

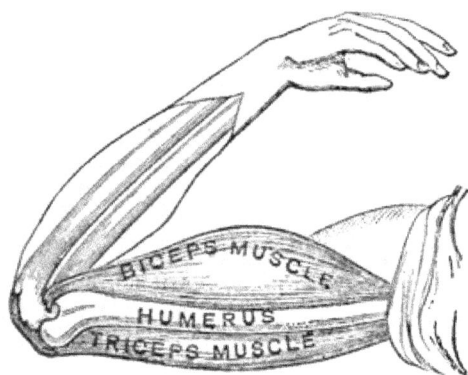

FIG. 25. — Biceps and Triceps Muscles.

did, the arm would simply fall into the straight position. A muscle at the back of the upper arm, called the triceps, the three-headed muscle, now contracts in obedience to the will, and by its contraction straightens the muscle.

**41. The Tendons.** — If we bend the leg or arm, and grasp the inside bend of the joint with the hand, we feel the motion of cords just beneath the skin. These are the **tendons,** or sinews, forming the tapering ends of muscles, which are fastened to the bone.

Tendons are white, glistening cords, like belts or straps, which connect the muscles with the bones. They

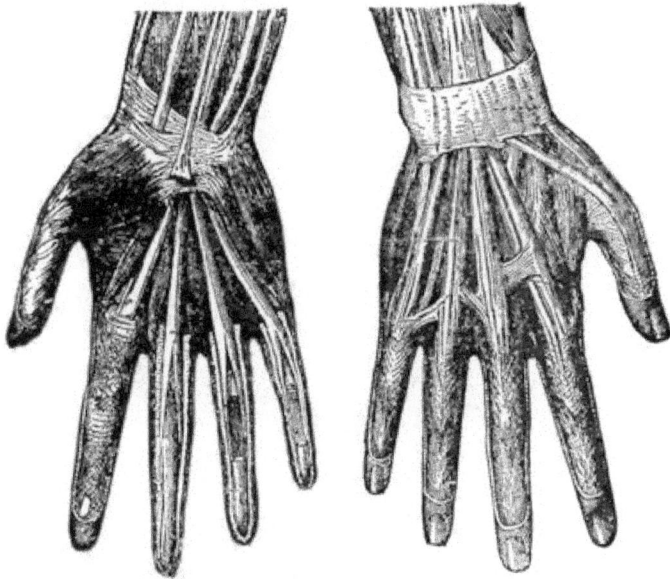

FIG. 26. — Muscles and Tendons of the Hand.

are very strong, but flexible. Children often amuse themselves by taking the leg of a fowl, a rabbit, or some small animal, and moving the toes by pulling a white cord in the leg. This cord is a tendon. Tendons are most numerous about the joints, especially the larger ones, like the knee and elbow. They save a great deal of space, and allow great freedom of movement, where muscles large enough to do the work would be exceedingly bulky and clumsy.

The longest and strongest tendon in the body is the tendon of Achilles.   The tendons of the muscles which bend the knee, and are felt in the bend of the joint, are commonly called the " hamstrings."

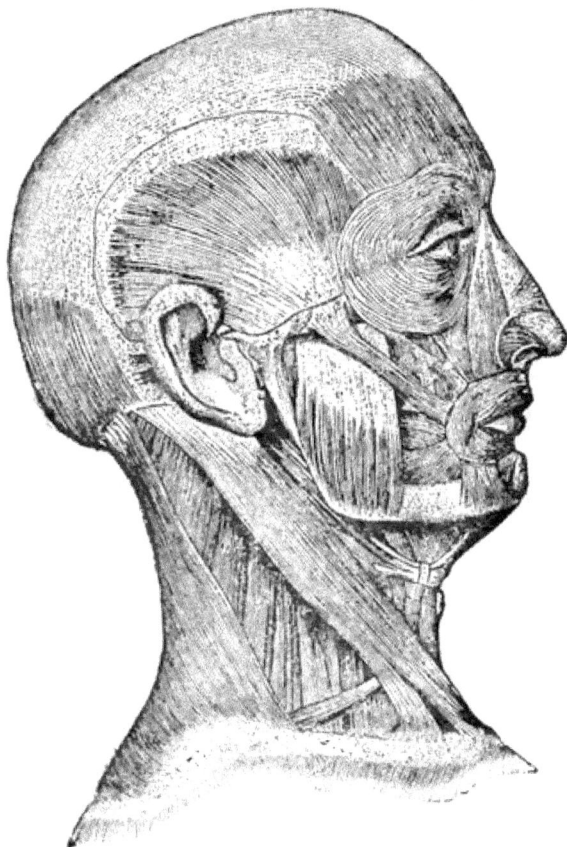

FIG. 27. — Muscles of the Face and Neck.

**42.  The Bony Levers of the Body.**—The word "lever" means simply a rigid bar, or rod, which moves about a certain fixed point, called the fulcrum.   We have all seen a man trying to raise a heavy stone with a crowbar. He places one end of the bar under the stone, and presses down the other end. In this case, the bar is the lever, and the ground under the bar the fulcrum. The pressure which the man applies is called the power, and the stone to be lifted is called the weight.

We have the same thing in a seesaw, where the plank is the lever, and the block on which it rests, the fulcrum; while the power and weight are first at one end, and then at the other. In a lever, then, we have three things to consider, — the fulcrum, the power, and the weight. Levers are divided into different classes, according to the position of these three things.

(1) *First Class.* — Levers of this class have the fulcrum in the middle.

(2) *Second Class.* — In this class, the fulcrum is at one end, the power at the other, and the weight in the middle.

(3) *Third Class.* — In this class, the fulcrum is at one end, the weight at the other, and the power between them.

In the various movements of the body, we have instances of all three kinds of levers.

The skull, as it nods backwards and forwards upon the atlas, is an instance of the first class of levers. It is like the seesaw we have just spoken of. The skull rests on the spinal column, which is the fulcrum; the muscle at the back of the neck is the power, and the front part of the skull is the weight. We have all seen a woman running a sewing-machine with her feet. In this case, the ankle-joint is the fulcrum; the tendon of Achilles, attached to the heel-bone, is the power; and the weight is at the toes.

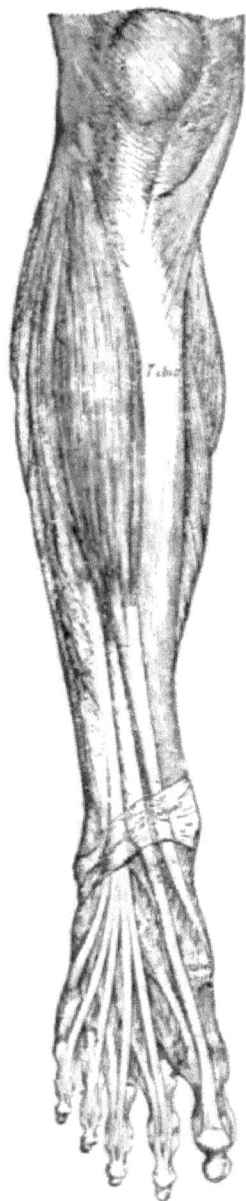

FIG. 28.

This is another instance of the first class of levers. When we stand on tiptoe, the power required to raise the body is applied to the heel-bone by its stout tendon. The toes are the fulcrum, and the bones of the foot support the weight. We have in this case a lever of the second class.

The third class of levers is the one which occurs most frequently in the body. Of this we have an example when we bend the fore-arm. The elbow is the fulcrum, the fore-arm and hand are the weight, and the biceps muscle, at its insertion in the radius, is the power which is thus placed between the weight and the fulcrum.

**43. A Few Important Muscles.** — There are about five hundred muscles in the human body, all necessary for performing the various movements and operations of this complicated machine, — our body. There are about fifty in each arm and hand. Six little muscles move each eyeball.

Muscles are of all kinds of shapes, — round and flat, long and short, fan-shaped, etc., — suited to the work they are to do. Some are large, like the great breast-muscles; while others are very small, like the tiny muscles which move the delicate little bones of the ear. Each muscle has its own name, given to it from its peculiar shape or size, or from the work it has to do. We chew our food by the help of two strong muscles, called the "chewing-muscles," which move the jaws. They are very large and strong in flesh-eating animals, like the lion and tiger.

Turn the head suddenly to one side, and the sharp edge of a long muscle is plainly seen and felt on each side of the neck: one end is tied to the skull, the other to the collar-bone. It acts to turn and brace the head. Inside the cheek is a flat muscle, called the trumpeter's muscle. It is largely developed in glass-blowers, and persons who play on wind-instruments. The muscle on each

FIG. 29. — Muscles of the Sole of the Foot, first layer.

side of the chest is fan-shaped, and powerfully developed in strong men.

The large, thick muscle covering the shoulder is thought to resemble the Greek letter $\Delta$ (delta). A military officer wears his epaulet over this muscle. The action of the two-headed muscle, which bends the fore-arm, and the three-headed, which straightens it, has been explained. A three-sided muscle covers the shoulder-blade, like a monk's hood, and helps move the shoulder. A very broad muscle in the back is the "climbing-muscle" of the body, and helps pull the arm backwards.

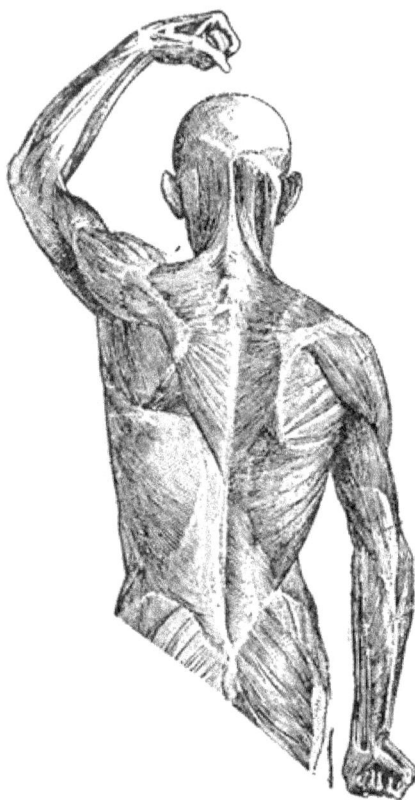

FIG. 30.—Some large Muscles of the Back.

Three huge muscles make up the greater portion of the fleshy mass in the lower part of the back. They move the thigh backwards, and help keep the body erect. The longest muscle in the body, called the "tailor's muscle," runs across the thigh in front. It helps us to cross the legs. Two strong muscles form the largest part of the calf of the leg. The tendons of these two muscles unite to form the tendon of Achilles. These muscles are used in walking, standing, dancing, and leaping. They are of great strength; because in raising the heel, they have to raise the weight of the body.

44. **Effect of Alcohol on the Muscles.** — We have learned that it is the nerve-force acting upon the muscles which makes them contract or relax. Now, if we drink more or less of alcoholic liquor, the muscles are acted upon in a peculiar way. The nerve-force that controls the muscles is weakened, and they show a lack of control.

The delicate movements which require the long training of certain muscles, as in handling fine tools and

**Fatty Decay due to Alcohol.** — All the important organs of the body have special work of some kind to do, and a peculiar power of doing it. They are made up of various forms of albuminoid matter endowed with living powers which vary with the different tissues. Thus, muscle is made up of fibres which have the power of contraction; and nerves, of delicate threads which carry nerve impulses.

Now, alcohol has the power of changing this vital property of the organs, someway related to albumen, into fatty or oil globules. This we may, for convenience, call fatty decay. Scientific men call it "fatty degeneration." It simply means that fat takes the place of the albuminoid matter. In brief, there is a decay in the living body of parts of our most important organs.

This decay occurs most frequently in the tissues of the heart and liver; but it may go on in the blood-vessels, in the muscles, in the nerves, and, indeed, in all the vital organs. Alcohol taken for some time often causes this fatty decay. Other poisons, as phosphorus, produce it more rapidly; but alcohol causes it in a much more general way. The structure of the organs thus diseased is weakened, and their function seriously impaired.

The heart, which is composed chiefly of muscle, is very liable to become diseased through the unhealthy accumulation of fat caused by beer-drinking.

doing gymnastic feats, cannot be made. A person may know the right way of making each movement, and may succeed, after a fashion, in clumsily doing it; but the trained muscles are no longer wholly under the control of the will. If alcohol enough is drunk, all muscular control may be lost, and deep breathing may be the only sign of life.

This same lack of control is shown in the act of speech. Each and every word we utter requires special movements of the muscles of the tongue, palate, and throat, all acting in harmony. After drinking alcohol, there is less control of the muscles: the reins are slackened, so to speak; words may be left out, cut short, or misplaced. According to the stage of intoxication, the words are clipped, stammered, or "mouthed," or "thick," from loss of control of the tongue.

The muscles that move the eyes do not act in harmony: hence the drunken man "sees double." The rapidity with which this loss of control is produced varies with the individual, with the kind of drink, the rate at which it is drunk, and many other circumstances.

**45. Effect of Alcohol on Muscular Strength.** — People have supposed that alcohol gave them strength for their work, and rested them when they were tired. In both cases this supposition is wrong. Instead of adding to strength or diminishing weariness, it only deadens the nerves and impairs the judgment. For when the strength is tested with a health-lift or other means, the drinker is found to be weaker after taking an alcoholic liquor than before. After the effect of the alcohol has passed off, the feeling of weariness is more intense than before, showing that the alcohol did not remove it, only concealed it.

A repetition of the drink may again create insensibility to the fatigued feeling, and the muscles may again obey the will; but only for a briefer time than before. In this way the man who could have put forth just as much strength in an emergency, and could have held out longer, accomplishes less work, abuses his muscles, and deludes his mind by resorting to alcoholic drinks. He has also the injurious effects of the alcohol on other parts of his body to contend with afterward.

Physiologists tell us that by the use of the microscope in the dissecting-room they find a notable difference between the muscles of the heavy, inactive beer-drinker and those of the total abstainer. The latter are firm, elastic, and of a bright red color. The former are pale, flabby, and often loaded with fat.[1] Tiny particles of fat collect in and around the muscle fibres, and crowd upon them, greatly reducing their strength.

**46. Effect of Tobacco on the Muscles.** — Tobacco causes relaxation of the muscles, and weakens the nerves which control muscular movement. The result is seen in the unsteady hand of the cigarette-smoker when he attempts to draw a straight line or do other nice work which requires precision of touch. And his loose, shuffling gait gives observers the impression that his muscles are too weak to hold his bones together.[2]

1 See Note 2, page 356.          2 See Note 3, page 358.

# CHAPTER IV

## PHYSICAL EXERCISE

**47. Why we need Physical Exercise.** — To be vigorous and healthy, every organ of the body must be used. Every organ and every tissue must act well its part to be of the best service. To keep the body in health, it is absolutely necessary that a certain amount of muscular action or **physical exercise** should be taken every day.

The reason for this is plain : the body is made up of certain tissues, such as bone, muscles, and nerves. These tissues consist of a countless number of little cells, as we have learned, every one of which is born, lives its brief moment, then dies, and is cast out as waste or dead matter. The ever-changing blood, in its ceaseless current, brings them their nutriment. The tissues take what is specially suited to their wants, and return as waste matter whatever has become used up.

Now, in every tissue, especially in the muscular tissue, this process is hastened by action. Exercise causes more frequent changes in the tissue-cells, and hence an increased flow of blood. The greater the amount of judicious exercise, the greater is the working-power of the individual. Muscular activity is, then, the chief agent in bringing about these wholesome tissue-changes.

Muscles increase in size and strength according to the use made of them. The blacksmith uses vigorously the muscles of his right arm day after day, hence their size and strength become well developed. Change his business for that of a clerk, and the once brawny arms soon become small and weak. Let a muscle be kept idle for some time, and it loses in bulk and vigor. Some of us have heard of certain people in India, who, as an act of worship, keep one arm raised above the head for many weeks. The muscles shrivel, and the arm dries up and becomes useless. If a leg is broken, and kept in splints for several weeks, the muscles become feeble and wasted. It is only after a great deal of exercise, that the long-idle limb regains its former size and vigor.

**48. Effect of Exercise on Various Organs.** — Not only are the muscles themselves benefited by exercise, but, by their action, many other tissues and organs become more vigorous. The heart beats more vigorously in order to carry more blood to the tissues. More oxygen is taken in by the lungs, and more heat is developed. Hence the skin, kidneys, and lungs have to do more work to get rid of the waste products.

Breathing is mainly done by muscular action : thus, exercise causes the lungs to draw in an extra amount of fresh air, and to get rid of more impure air. Again, exercise stimulates the muscles which control the organs of digestion, thus giving a good appetite by creating a demand for food.

In brief, muscular exercise is needed to keep the whole machinery of the body in good working order.

**49. Amount of Physical Exercise.** — Too much exercise, as well as too little, is a fruitful cause of ill-health. Exercise followed by fatigue, day after day, only does harm ; while judicious exercise with suitable rest is of great benefit. Hence the amount necessary to keep the body in the best order is a most important and practical question.

It is laid down as a pretty safe rule, that a person of average height and weight, engaged in study or any other in-door or inactive business, should have an amount of exercise equivalent to a daily walk of five miles along a level road. Growing children, as a rule, take more exercise than this; while most men in the prime of life, working in-door, take nothing like this amount ; and many women take less. Of course, if one's daily

**Health and strength contrasted.** — "Health and strength are not synonymous terms. A person may have great strength in his limbs, or in certain muscles about the body, but really not have good health. It is altogether a mistaken idea to suppose that physical exercises have for their sole object the attainment of strength. There are other tissues and organs in the human system besides the muscular ; and the healthy action of the lungs and the stomach is far more important than great strength in the arms, legs, or the back. It is here, in this general exer-cise of all the muscles and parts of the body, that a well-regulated system of gymnastics has its great excellence. It aims to produce just that development of the human system upon which good health is permanently based, described by a distinguished writer as follows: 'Health is the uniform and regular performance of all the functions of the body, arising from the harmonious action of all its parts,' — a physical condition implying that all are sound, well-fitting, and well-matched. Some minds do not look far enough into life to see this distinction, or to value it if seen; they fix their eyes longingly upon *strength*, — upon strength *now*, — and seemingly care not for the power to work long, to work well, to work successfully hereafter, which is *health*." — DR. NATHAN ALLEN *on Physical Culture.*

work is active and out-door, no additional exercise is really necessary. Exercise may be varied in many ways, — the more ways the better. But, for the most part, it should always be taken in the open air.

**50. Time for Exercise.** — The best time to take exercise is about two hours after a meal. The body is the weakest before breakfast. It is not best to do hard work, or take severe exercise, before this meal. Those who go to work or study before breakfast, should first eat half a slice of bread, or a cracker, or even drink a glass of milk, — just enough to "stay the stomach," and save the feeling of faintness, or "sinking" at the stomach.

Just after a full meal, the stomach is busily doing its duty. Hence exercise at this time is apt to stop its action, and result sooner or later in dyspepsia. The evening is not the best time for exercise, because the body is tired after the labor of the day. It is useless to fix any exact rule. Ordinary work, or moderate exercise, as walking, is beneficial almost any time, except just after a full meal.

**51. Different Kinds of Physical Exercise.** — The kind of exercise it is best to take depends very much upon one's daily occupation. Persons who sit at desks, stand at counters, or work in close rooms, as clerks, teachers, tailors, printers, etc., are prone to diseases due to lack of bodily exercise and to foul air. Every person should know, or be taught to know, his own needs and his own dangers.

Almost every person should do some work, or take some exercise, with both body and mind every day. To get and to keep vigorous health, it is not at all necessary to increase the size of the muscles very much, or do great feats of strength. Walking is the best of all exercises. It takes us into the open air and bright sunlight. It puts new life into many important muscles of the chest, abdomen, and limbs. With a brisk walk every day, taking care to keep warm and dry, no one need suffer from lack of proper exercise. Running, leaping, climbing, and other vigorous sports are well enough, especially for children, if they are not kept up too long, and do not cause fatigue.

Violent sports, such as base-ball and foot-ball, are severe exercises, and occasionally dangerous. Rowing is admirably suited to most persons of either sex. Horse-back and bicycle riding, coasting, swimming, tennis, and skating are important helps to increase the bodily vigor. Certain sports also tend to beget self-reliance, coolness in danger, and a certain dignity and grace of person. There is hardly any one kind of exercise, which, taken alone, is able to give even tolerable development of all the muscles. Hence light gymnastic exercises are cheap and convenient means to develop muscles not used in work and games.

Growing children should be trained every day at home or in school in the use of light wooden dumbbells, light clubs or wands. A daily exercise of ten minutes will do much to develop feeble and narrow chests, to check the tendency to curvature of the spine and round shoulders so common with girls, and to give muscular strength and vigor to all parts of the body.

**52. Beneficial Effect of Physical Exercises in Schools.** —The measure of the health of young people depends to a great degree upon a proper amount of physical training. Hence pupils should have some sort of physical exercises provided for them in the schools. This is especially true in large towns and cities, where there is less opportunity for out-door games.

Such exercises, whether of one kind or another, should form a part of the regular course of study. The object aimed at should be the promotion of health and power of work, rather than the development of muscle or the doing of feats of agility and strength.

Such exercises, if of real value, increase the breathing-power, and quicken the action of the heart. They fill the arteries with pure blood, and distribute it, with increased energy, to all the tissues and organs of the body, stimulating them to renewed activity. They reinvigorate the whole system, and at the same time furnish what is very important, — a pleasant recreation.

To obtain the greatest amount of good from all movements calculated to improve the physical and mental strength of young people, it is highly important to remember that pupils should be interested, and made to feel that these exercises are a recreation instead of a task.

**53. The Need of Physical Training in our School System.** —Nearly one-half of the working hours of every pupil is spent in the schoolroom itself, or in the preparation for the work there done. The conditions for the best physical development are rarely reached in our schools.

Impure air, lack of proper ventilation, faulty positions long continued, and other conditions unfavorable to health and normal development, demand a rest for the over-tired muscles and the over-taxed nerves. Hence the connection between physical training and school education is obvious.

To the natural varieties of games and exercises used by boys — base-ball, foot-ball, bicycle, croquet, tennis, rowing — should be added systematic instruction in the schools. This should be especially directed with a view to develop the neglected and weak parts, and to add to the symmetry of the whole.

The real benefit of systematic physical exercises in the schools, however, does not depend upon any large amount of development that they afford. They should be adapted to a harmonious development of the whole physical system, so as to result in a perfect control of every muscle, producing grace and freedom of movement, as well as health and vigor.

**Importance of Physical Development.** — " In early boyhood and youth nothing can replace the active sports so much enjoyed at this period; and while no needless restrictions should be placed upon them, consideration should be paid to the amount, and especially to the character, of the games pursued by delicate youth. For these it would be better to develop the weakened parts by means of systematic physical exercises, by short excursions into the country, and by lighter sports.

"Children who are taught at an early age to be obedient seem to enjoy more thoroughly such exercises as combine discipline with rhythmic movements; and consequently, the older the child, the more important it is to adopt a system of calisthenics, or light drill, or games that combine gymnastics with rhythmic sounds and periods of rest."—DR. JOHN M. KEATING on *Physical Development* in Pepper's *Cyclopædia of the Diseases of Children.*

Simple exercises, without any apparatus, practised a few times every day for five or ten minutes at a time, do a great deal of good. They relax the tension of body and mind, and introduce an element of pleasure into the routine of school life. These simple but vigorous and systematic movements are of great use in schools of the lower grades, and are widely used at present with a real benefit.

**54. The Various Systems of Gymnastics.**— It is not the purpose of this book to discuss in detail the merits and defects of the various systems of gymnastics now in general use in this country. They all have strong points — and their weak ones.

The present revival of popular interest in all that pertains to physical culture has stimulated our leading educators to renewed activity in urging a more systematic use of gymnastic exercises in our schools.

FIG. 31.—Showing the "wing" position (arms resembling the wings of a bird) in the Swedish drill.

We have sketched briefly in the following sections a few of the essential facts or principles that give character and utility to the two great systems (the Swedish and German) of gymnastic training. Upon these two systems are based many of the valuable gymnastic movements now in common use in our schools.

**55. The Swedish System of Gymnastics.** — The Swedish system of physical exercises as arranged by Ling, and modified to some extent by the teachers of this method since his time, is generally regarded as having very important merits. The main principle which underlies all the movements of this system is to impart invigorating forces to the vital organs.

Every movement is carried out with deliberation and mental attention, and designated by a phrase which can be used as a *word of command.* This word of command is the only method, the advocates of this system claim, which enables the pupil to concentrate his mind on one thing at a time, that thing being his own movement.

Words of command have other advantages. They teach the pupil to think quickly, to act as quickly, and to do a thing in the shortest possible time.

Again, in the Swedish system *regularity of method* is adhered to very strictly. The exercises beginning by the very simplest, gradually be-

FIG. 32. — Showing the "rest" position (on account of its restful feeling) in the Swedish drill.

come stronger and more complicated as the pupils develop. The slightest change of position — even the turning of a hand — has its recognized influence in the progression.

The exercises of the Swedish system can be practised in the school-room, most of them by the class in con-cert, and equally by boys and girls. The method does not approve of music, and *requires no apparatus.* The exercises, not the apparatus, constitute the system.

The movements can be made anywhere where there is sufficient floor-space to stand on and sufficient oxygen in the air. The apparatus can be easily adapted to the apparatus belonging to other systems, or to such simple means as ordinary chairs and desks, and other furniture. Though apparatus is desirable, it is not absolutely necessary for good physical development, especially in gymnastics for children.[1]

[1] " The exercises of the Swedish system are chosen according to their *gymnastic value*, which quality depends on how the movement combines the utmost effect on the body with simplicity and beauty of performance. Only such exercises are used whose local and general effects are fairly well known and proved to be needed by the body. Not only the needs of the individual, but his abilities as well, are to be taken into consideration; and for that reason the teacher must know how to vary the exercises according to the degree of physical culture possessed by the pupil. The movement should have its developing effects in a short time; it should be simple, so that every pupil can do it fairly well; and it should have beauty of execution, according to each one's ability." — BARON NILS POSSE, M.G.

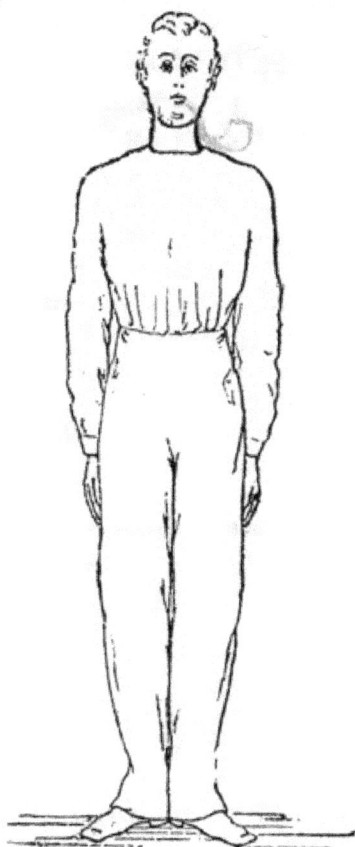

FIG. 33. — Standing position in the Swedish system, ready for gymnastic drill.

Note. — For a most comprehensive work on the Swedish system, the teacher is referred to the " Swedish System of Educational Gymnastics," with 264 illustrations, by Baron Nils Posse, M. G., Director of the Posse Gymnasium, Boston. A small manual for teachers, called " Handbook of School Gymnastics of the Swedish Systems," by Baron Nils Posse, has been recently published.

## 56. The German System of School Gymnastics. —

The German system of gymnastics ranks high among all the different systems. It has been built up during almost a century by men of science. It is practised in classes at school by hundreds at the same time, and by single persons as home exercises.

**The Benefits derived from a Modern Gymnasium.** — "I think it will be admitted by all thoughtful persons that one-half the battle for mental education has been won when you arouse in a boy a genuine love for learning. So one-half the struggle for physical training has been won when he can be induced to take a genuine interest in his bodily condition, — to want to remedy its defects, and to pride himself on the purity of his skin, the firmness of his muscles, and the uprightness of his figure.

"Whether the young man chooses afterwards to use the gymnasium, to run, to row, to play ball, or to saw wood, for the purpose of improving his physical condition, matters little, provided he accomplishes that object.

"The **modern gymnasium,** however, offers facilities for building up the body that are not excelled by any other system of exercise. The introduction of the new developing appliances has opened up the possibility of the gymnasium to thousands to whom it was formerly an institution of doubtful value.

"The student is no longer compelled to compete with others in the performance of feats that are distasteful to him. He can now compete with himself — that is, with his own physical condition — from week to week, and from month to month. If he is not strong enough to lift his own weight, the apparatus can be adjusted to a weight he can lift. If he is weak in the chest or the back, he can spend his time and energy in strengthening those parts without fear of strain or injury.

"In fact, he can work for an hour, going from one piece of apparatus to another, keeping always within the circuit of his capacity, and adding slowly and surely to his general strength and powers of endurance. If the heart is weak, the lung-capacity small, the liver sluggish, the circulation feeble, or the nervous system impaired, etc., special forms of exercise can be prescribed to meet these conditions." — DR. D. A. SARGENT, *Director of the Hemenway Gymnasium, Harvard University.*

The instruction begins with simple and easy movements, and gradually proceeds to those more difficult. The apparatus used in school practice is not complicated or expensive. It may be omitted altogether if the necessary room for climbing-poles, ladders, and some light apparatus cannot be provided for. An almost endless variety of simple and complicated free exercises may be made, with or without the common hand apparatus, as wands, dumb-bells, and Indian clubs.

In the German schools the lessons begin regularly with a series of free and order exercises. Every scholar has to take part in them. The rhythmical order in which they are produced calls for absolute attention, and allows no backwardness. Class exercises on apparatus follow the free exercises. A change of apparatus takes place, and then the lesson ends with some exercises left to individual inclination.

FIG. 34. — A schoolgirl in gymnastic costume.

**57. The System needed for our Schools.** — The various systems of physical exercises as practised in Europe have proved acceptable in this country, but with many important modifications. American educators have ideas of their own on

physical culture. We are slow to adopt the methods of other countries unless they are modified to suit our special demands.

Many of our ablest teachers of physical exercises believe in music in gymnastic training. They claim that many of the lighter exercises can be better executed with music.

It is also maintained that it is no relief from the strain on the brain to go from the study of arithmetic to the performance of physical exercises that demand close attention to the teacher who gives the signal for the gymnastic exercises.

In brief, the ideal system adopted for our schools must be an eclectic one. It will assimilate all that is best in every known method. It is not the system itself which must bear the test of criticism, but the way in which the system is taught.

**58. Effect of Alcoholic Liquors and Tobacco on Physical Exercise.** — The main object of physical exercise is to get our bodies into such a condition, and to keep them in that condition, whereby the average amount of working-power can be utilized at any time without harm to the bodily health. To keep up this amount of physical power and endurance we must be obedient to certain great laws of health.

One of these laws which never can be violated with impunity is that which forbids the use of alcoholic liquors and tobacco. Strong drink and tobacco will put to naught the most elaborate and costly system of physical training.

Those who train athletes, base-ball and foot-ball

players, oarsmen, and all others who take part in severe physical contests, understand this, and rigidly forbid their men to touch a drop of alcoholic drink, or even to smoke or chew tobacco. Experience has proved beyond all doubt that strong drink is a positive injury, either when men are in training for or undergoing contests demanding long-continued physical endurance.

The same law holds good in the ordinary physical exercises of every-day life. Alcohol and tobacco act as poisons to the nerve-force which controls the muscles, and thus lessen the amount of muscular power and endurance.

The demands of modern life call for a *sound* body rather than a *strong* body. Neither is possible to those who indulge in alcoholic drink or tobacco.

Note. — In the Appendix may be found a series of simple gymnastic exercises, systematically arranged, which may serve as the basis of other and more extended exercises suitable for every-day use in the schoolroom.

# CHAPTER V

## FOOD AND DRINK

**59. Work, Waste, and Repair.** — The body is in some ways — as we have learned in the Introduction — very much like a steam-engine. The bones and muscles correspond to the machinery. The motive-power is furnished by the **food** we eat. The food is to us what coal and wood are to the engine.

Like the locomotive, our bodies move about, and are warm, because a slow fire is always burning in them. This fire, like that of the engine, needs fresh fuel from time to time. The food we eat becomes part and parcel of the body, and it is the *whole* body that is slowly burning. In order to keep up this bodily burning, we must have oxygen. Without fuel and air, the fire in the engine will go out.

So it is with the fire in the body. Without food and air, the combustion of our bodies would soon flag, and we would soon die for want of them. When coal or wood is burned, we get ashes. When food is burned in the body, we get, not exactly ashes, but something like it, which is got rid of as useless substances.

Again, the steam-engine wears out, and needs repair from time to time. So it is with our bodies. There is wear and tear constantly in every part. Every beat of the heart, every contraction of a muscle, and even our very thoughts, lead to waste. Every time we think, look, speak, or move, we do so at the expense of some minute portion of the tissues of the body.

60. **Why we need Food.** — To make good this waste, there must be something taken into the body, or it soon suffers. The body is constantly wearing away, and must be constantly built up again. Hence we see the necessity for food. The food, too, must contain the same things that the body has lost.

The blood is the stream which carries this material for restoring the waste tissues of the body. Indeed, we may say that the blood is that material itself. It is simply the food we have eaten, in another form. Food must be of such a kind, and in such a quantity, as to supply an amount of nutriment to the tissues equal to the waste which takes place.

During early life, when the body is growing rapidly, more of tissue-food is needed than in adult life, when repair alone is called for. When the body has been wasted by disease, as typhoid fever for instance, much tissue-food is needed to repair the waste, and to restore the muscles to their former size and vigor.

If but little or no food be taken, or if it be not of the right sort, the body slowly loses in weight. If we try to do without food, we grow chilly and cold, feeble, faint, and too weak to move. A point will at length be reached at which death from starvation must take place. The bodily fire has gone out, and nothing but cold ashes is left.

**61. Different Classes of Food.** — For convenience let us divide foods, or foodstuffs, in four great classes : —

I. Nitrogenous **Foods, or Albumens.**

II. Starches and **Sugars.**

III. Oil and **Fats.**

IV. Inorganic or **Mineral Foods.**

**62. Nitrogenous Foods, or Albumens.** — These foods, sometimes called flesh-forming foods, contain, as their name implies, all the materials requisite for restoring the tissues. They contain all four elements, — nitrogen, carbon, oxygen, and hydrogen. They are especially rich, however, in nitrogen, which forms an essential part of the animal tissues.

These foods are called albumens, because they contain a certain white substance known as albumen, familiar to all as the white of an egg. Lean meat; the cheesy part, or curd, of milk; pease and beans, — are

---

**The Wear and Tear of the Body.** — "In the physical life of man there is scarcely such a thing as rest : the numberless organs and tissues which compose his frame are undergoing perpetual change, and in the exercise of the function of each, some part of it is destroyed. Thus, we cannot think, feel, or move without wasting some proportion, great or small, according to the energy of the act, of the apparatuses concerned, — such as brain, nerve, or muscles. Now, this waste-product cannot remain in its original situation, where it would not only be useless dross, but also obstructive and injurious. Such old material is being daily removed from our bodies to the average amount of three or more pounds ; and that an equal quantity of new shall take its place is the first principle of alimentation. This tissue-change is so complete, that not a particle of our present body will be ours a short time hence ; and we will be, as I have lately seen it phrased, like the knife which, after having had several new blades, and at least one new handle, was still the same old knife to its owner." — MAPOTHER'S *Lectures on Public Health.*

rich in albumen. Wheat, barley, oats, and maize also contain albumen.

A well-known group of plants with seeds enclosed in pods, or legumes, hence called "legumens," is rich in a nitrogenous substance, resembling the albumen of eggs and the gluten of flour. Pease, beans, and other vegetables of this group, are nutritious on account of this flesh-forming material.

**63. Starches and Sugars.** — The starches and sugars contain carbon, hydrogen, and oxygen, but no nitrogen. They are sometimes called the "sugar hydrocarbons." As their proportions of hydrogen and oxygen are the same as in water, we may consider them as carbon or charcoal dissolved in water.

This class of foodstuffs forms a large proportion of all those plants which are generally used as food. Wheat, barley, oats, rye, rice, maize, tapioca, arrowroot, sago, potatoes, etc., are rich in starch. In fact, starch stands first in importance among the various vegetable foods. Starch in its natural state would be useless as food, because it is insoluble. It is only after it has been acted upon by the digestive fluids, and converted into sugar, that it becomes soluble, and is capable of being taken up by the blood.

The sugars themselves come from the fruits and vegetables, honey, and milk. The sugars serve as a kind of fuel, which is burned up by the oxygen in the body. The combustion of these fuel-foods develops and keeps up the necessary heat of the body, and the vital energy or strength which enables us to perform our various duties.

**64. Oil and Fats.** — These foods contain carbon, hydrogen, and oxygen. They are sometimes called the "fatty hydrocarbons." This class of foods is derived from the fat of meat, eggs, butter, and milk, also from the various oils. Most of the breadstuffs contain more or less of fat or oil. The oils and fat supply about twice the amount of energy, whether manifested as heat or motor force, as that afforded by the combustion of an equal amount of starch or sugar.

**65. Inorganic Substances, or Mineral Foods.** — Organic substances, such as those included in the food-stuffs just described, are derived from living forms, as animals and vegetables. Besides these, the body demands a certain amount of inorganic substances, derived from the mineral kingdom. These are water, salt, iron, lime, magnesia, phosphorus, and potash. Except water and common salt, these substances generally enter the body only in combination with other foodstuffs.

**66. Extra Foods, or Appetizers.** — A vigorous appetite requires little else besides salt for an appetizer. Salt has been and is universally used by all animals and peoples. Salt has always been the symbol of life, hospitality, and wisdom. Animals will travel for miles to reach salt-licks, while men have risked their lives to get even a taste of salt. Pepper, mustard, ginger, and other heating appetizers, are used to give flavor to tasteless foods, and to stimulate the sense of taste, and to increase the flow of saliva and gastric juice. So long as the "appetite comes with eating," and ordinary food is enjoyed, the use of these stimulating appetizers is unnecessary, and becomes simply a matter of habit.

**67. Different Articles of Diet. Vegetable Food. —**
The most important food of this class is bread, "the staff
of life." There is no single food in the world which meets
so many necessary wants of the body. Bread is made
from the flour of many kinds of seeds, such as oats, rye,
Indian corn, etc. But in this country it is nearly
always made from wheat.

Wheat-flour gives us starch, sugar, gluten, — a form of
albuminous food. Hence wheat has nearly everything
to support life except fat. When we eat bread and
butter, we have nearly a perfect food. Corn-meal is rich
in nitrogen, and has much oily matter. It is highly
nutritious and a cheap article of food. Oatmeal is
richer than flour in nitrogen and fat, and is therefore
more nutritious. Rice, though rich in carbon, is one of
the least nutritious of all the cereals. Pease and beans
contain more nitrogen than any of the cereals, and are
as rich in carbon as wheat-flour.

The common, or Irish, potato is a most important
article of diet. Although it is more than two-thirds
water, and has little nutriment, yet it is easily digested,
— a cheap and economical article of diet. Ripe fruits,
such as apples, pears, peaches, melons, grapes, oranges,
etc., though not of much nutritious value, are prized
for their agreeable flavor. Sugar and molasses are both
largely used in cooking. Their nutritious value is
about the same as that of starch.

**68. Animal Food. —** The first place must be given to
milk, for it is the food of all others which gives us
nourishment in the simplest and best form. It contains a large quantity of water, caseine, together with

sugar and fat. Eggs have a large amount of nutriment in a small space. Two-thirds of an egg is water, the rest albumen and fat. Eggs are cooked in many ways, and are generally easily digested.

Meats, for the most part, consist of the muscles of the various animals. The most common are beef, mutton, lamb, veal, and pork. Meat is rich in albumen, and has more or less of fat. It is a most common and important article of diet, and, as a whole, is easily digested, except, perhaps, veal and pork. Fish is at once a cheap, abundant, and nourishing food. Poultry is a useful, light article of food, especially for the sick and feeble. The flesh is easy to digest, and gives a deal of nourishment.

**69. Mineral Food.** — Water is a mineral food, which will be described hereafter. There is about a half a pound of common salt in the body at any one time. But we are continually losing it. Tears, we know, contain salt; and it is also found in the sweat. Many people think they do not eat any common salt, because they do not take it by itself; but they forget that many of the foods they eat, such as bread and meat, have a little of it.

The salts of potash are chiefly found in the vegetables we eat, especially lettuce. These salts help purify the blood. The salts of lime make the bones hard and strong. Iron is contained in very small quantities in many of the foods we eat. It helps to make good blood.

**70. Natural Drink. Pure Water.** — Drink is of just as much importance as food. Every one knows what happens to plants when they are deprived of their drink. They first droop, and then, soon afterwards, wither and die. So also it is with all animals. If they are deprived of their drink, they also droop, and at last die. Animals may be dried away into death as certainly as plants.

Drink is the agent which has been provided by nature to wash the food into and through the living tissue of bodies. The drink provided by nature for all living creatures is **pure water.** Wild animals, as well as domestic, often take no other drink but water.

**The Value of Salt.** — In most countries, salt is "cheap as dirt;" while in parts of Africa it is worth its weight in gold. In that land brothers will barter their sisters, husbands their wives, and parents their children, for salt. On the Gold Coast of Africa a handful of salt is the most valuable thing upon earth after gold, and will purchase a slave or two. Travellers tell us that with certain tribes the use of salt is such a luxury that to say of a man, " He flavors his food with salt," is to imply that he is rich; and children will suck a piece of rock-salt as if it were sugar. In India a large revenue is raised from the salt-tax, salt being something which even the poorest native will buy. In old times untold tortures were inflicted upon prisoners in Holland, by feeding them on bread alone, made without salt. No stronger mark of respect or affection can be shown in some countries than the sending of salt from the tables of the rich to their poorer friends. In the Book of Leviticus it is expressly commanded, as one of the ordinances of Moses, that every oblation of meat upon the altar shall be seasoned with salt, without lacking; and hence it is called the Salt of the Covenant of God. The Greeks and Romans also used salt in their sacrificial cakes. Everywhere, and almost always, indeed, salt has been regarded as emblematical of wisdom, wit, and immortality. To taste a man's salt, was to be bound by the rites of hospitality; and no oath was more solemn than that which was sworn upon bread and salt. To sprinkle the meat with salt was to drive away the Devil; and to this day, among the superstitious, nothing is more unlucky than to spill the salt.

Thousands of human beings, following the example of the lower animals, drink nothing but water, but yet toil long and hard, and fulfil all the duties of a reasonable and intelligent existence.

In brief, it is clearly evident that pure water is the only drink that is absolutely essential for all the purposes of life.

Now, it is plain that if we take a quart or more of drink every day, nearly an equal amount must be thrown out daily from the body. This is really the fact. Some portion of it steams away with the breath. More of it passes through the pores of the skin, and still more is drained away through the kidneys. Water thus drains off a great deal of waste and refuse matter that the body must get rid of. Water therefore drains waste matters out of the body at the same time it washes nourishing foods into it.

To be suitable for diet, water should be clear, without color, with little or no taste or smell, and free from any great amount of animal or vegetable matter. Real pure water does not occur in nature. Rain-water is the purest if properly collected. Well and spring water, and that brought into towns from some distant pond or river, all contain more or less mineral matter. The various mineral-spring waters, such as those at Saratoga, popularly used as medicines, are highly charged with mineral substances.

**71. Artificial Drinks.** — Our great natural drink, as we have learned, is pure water. Milk is also a natural drink. We have already learned some things about milk and water. Now, man has always contrived many ways to flavor his drink in some way or other. The greater portion of almost every drink is water; but, in various ways, other substances are mixed with the water, to give it a pleasant taste. Such drinks are artificial. Tea, coffee, and cocoa are the more common artificial drinks.

**Tea** was first brought to England about two hundred and fifty years ago, but it did not come into general use. The tea-plant is chiefly cultivated in China. When boiling water is poured on the dried leaves, a substance which chemists call an "active principle," and known as *theine*, is dissolved. The liquid so obtained is called a "decoction," and makes the common drink known as tea.

**Coffee** comes to us mainly from the West Indies, Arabia, South America, and the East Indies. When boiling water is poured on ground coffee, it dissolves a substance called *caffeine*, very much like the theine of tea.

The **cocoa**, or chocolate, tree is a native of tropical America. It has an active principle which resembles theine and caffeine.

Chocolate is cocoa ground up with sugar and certain spices.

One of the most common of all artificial drinks used in this country is **ice-water.** The temptation to drink freely of it in hot weather is rarely resisted. Ice-water quenches the thirst only for a moment. It weakens the strongest stomach after a time, and is thus a common cause of dyspepsia. If one must drink it, sip a little slowly.

The most common of all artificial drinks are **alcoholic liquors.**

In the next chapter we shall learn a great deal about their origin and nature.

**72. The Effect of Drinking Tea and Coffee.**— There is rarely any necessity for one to drink tea or coffee. They have little or no value as foods. Some persons cannot drink even a single cup of coffee or tea without feeling the worse for it : headache, indigestion, heartburn, wakefulness at night, and constipation are the most common after-effects. Strong tea should never be used.

Hard-working women and others, from choice or necessity, too often make their meals of dry toast and several cups of strong tea. Drank in excess, tea may weaken the action of the heart, and produce the peculiar beating, after much exertion, known as "palpitation." Hence we have "the tea-drinker's heart." Coffee and tea are unsuitable articles of diet for growing children.

Drinking strong coffee often gives a muddy look to the complexion of young folks ; much tea-drinking imparts a parchment appearance to the skin of young women. Doubtless all of us would be in better health if we never drank another drop of tea or coffee.

## CHAPTER VI

### ORIGIN AND NATURE OF FERMENTED DRINKS

73. **Decay a Law of Nature.** — If one is thirsty and cannot get water to drink, his thirst can be greatly relieved by juicy fruits. Fruit-juices are composed largely of water, sweetened with sugar, which nature prepares in them as the fruit ripens, and flavors each according to its kind. Such juices, as they come fresh from the fruit, are healthful.

But suppose the fruit is crushed, and the juice is drawn off and left standing in air of ordinary temperature. How long will it remain a healthful drink? Only a very few hours.

All animal and vegetable matter is composed of various simple substances, gases, liquids, and solids. Nature has provided that when plant or animal matter ceases to live, these different substances composing it shall be set free for use again in forming new combinations. When we see meat spoiling, bread or cheese moulding, fruit rotting, or sauce turning sour, we see simply an example of this provision.[1]

The "working," or fermenting, of sweet fruit, plant, or other vegetable juices, which takes place very soon after they are pressed out, is another example of the same wise provision. This last process is not accompanied by foul-smelling odors, as are most of the others, but by a peculiar bubbling of the liquid caused by the

[1] See Note 4, page 358.

escaping gases; hence the name "fermentation," from the Latin meaning "to boil."

**Fermentation** in its widest sense includes the changes going on in the putrefying meat, the moulding cheese, and the rotting fruit, as well as in the fermenting fruit-juice. They are all forms of decomposition that set free the simpler substances composing animal and vegetable matter. For all these one law holds good. It is this : —

FERMENTATION ENTIRELY CHANGES THE NATURE OF THE SUBSTANCE FERMENTED.

**74. The Cause of Decay.** — What causes all these various processes of decomposition ? Are plants and animals so constructed that when they have served their purpose they fall to pieces of themselves, or from the action of the air, as was once supposed ? No : the microscope has introduced to us whole tribes and families of minute living forms whose special work is to cause this decomposition of the dead bodies of their neighbors.

Were it not for these the whole surface of the earth would in time be covered with dead trees and other lifeless bodies, and there would be no simple substances left out of which to build up new ones.

Among these minute living forms are some that do not even wait for a plant or animal to die before attacking it. Such are the disease germs invisible to the naked eye, lurking in the air we breathe and in the water we drink, ready to cause sickness if given a lodgement.

Many kinds of these have been studied and described and grouped under the family name of **Bacteria,** the lowest of the plant-families.

The decay of meat and the souring of milk are caused by minute plant-cells belonging to this family. To another family low down in plant-life, viz., the fungi, belong the moulds that spoil our bread and cheese, and cause our sauce to sour and our fruit to rot.[1]

75. **Ferments and Their Work.** — To the fungi family also belong the class of microscopic plants called **Ferments** that cause the fermentation of sweet fruit, plant, or other vegetable juices. As long as fruit-juice remains inside the fruit, the ferment germs floating in the air, and even resting on the skins and stems of the fruit, do not enter it.[2]

When apples are ground to make cider, or grape-juice is pressed out to make wine, the ferments resting on their surfaces are washed into the expressed juice. Each ferment rapidly produces others, unless the liquid is kept at a very cold temperature, and all begin at once to break up the sugar of the juice.[3]

Two substances set free by this decomposition of the sugar are carbonic acid gas, which may be seen coming

---

[1] See Note 5, page 359.

[2] "Fermentation is the consequence of a development of vegetable cells, the germs of which do not exist in the saccharine juices within the fruits." — PASTEUR'S *Studies in Fermentation.*

[3] "In the fermentation of natural saccharine juices, which, especially when acid, undergo a decided alcoholic fermentation, the ferments originate in certain germ-cells which are spread in the form of minute spherical bodies of a yellow or brown color, isolated, or in groups, over the exterior surface of the epidermis of the plant, and which are gifted with an extraordinary power of budding with ease and rapidity in fermentable liquids." — PASTEUR'S *Studies in Fermentation.*

up out of the liquid in tiny bubbles, and alcohol, a poison which remains in the liquid.

The whole nature of the fruit-juice, which was wholesome and beneficial before this fermentation took place, is now changed by the presence of the alcohol. It becomes a poisonous liquid, called cider if made from apple-juice; wine if made from the juice of grapes, berries, currants, and other fruits.

If man should then leave this fermented liquid to itself, other kinds of ferments would enter it, and carry on the process of decomposition by setting up other kinds of fermentations, each after its own kind. For every fermentation, it must be remembered, has its own special ferment.

Among the minute plant-cells belonging to the bacteria family is one whose special work is to change the alcohol of a fermented liquid to acetic acid. Thus the fermented juice becomes vinegar, which contains no alcohol. Left again to itself, other forms of the bacteria enter the vinegar, causing still further changes, until finally the water is all evaporated and nothing but a little earthy matter is left. The work of decomposition is then finished.

But man interferes. He prevents, if possible, the microbe of vinegar from entering the fermented fruit-juice that he may use it for a drink. In many cases he is ignorant of the nature of the poison which the ferments produced by destroying the sugar.

76. **Alcohol as a Poison.** — Any substance capable, when absorbed into the blood, of injuring health or destroying life, is a **Poison.** Alcohol is capable of destroy-

ing life when taken in sufficient quantities, as has been proved by numerous instances of death following the drinking of spirits on a wager, or a draught of brandy or gin taken in ignorance by a child.

Alcohol is also capable of injuring health, as is proved over and over again every day. Few people have not known of one or more cases of loss of health from this cause, while physicians are meeting them constantly.[1]

Remember this:

ALCOHOL IS A POISON.

It is classed as such in standard treatises on poisons, in

**Alcohol a Poison.** — " Alcohol is universally ranked among poisons by physiologists, chemists, physicians, toxicologists, and all who have experimented, studied, and written upon the subject, and who, therefore, best understand it. It is not necessary to the action of poisons that they be always swallowed in fatal doses." — PROFESSOR W. J. YOUMANS.

Is alcohol a poison? I reply, yes. It answers to the description of a poison. It possesses an inherent, deleterious property, which, when introduced into the system, is capable of destroying life, and it has its place with arsenic, belladonna, prussic acid, opium, etc. In its effects upon the living system alcohol is first an irritant, and afterward, when it has entered the circulation, it becomes a narcotic. Were alcohol an irritant only, a man would as soon poison himself with arsenic. The narcotic element is the Siren that leads him on to ruin and to death." — DR. WILLARD PARKER.

" Alcohol is classed among the poisons by medical writers on poisons. I do not know of an exception among physicians. It is ranked among the poisons from its effects on the body analogous to those of the other poisons. What is said of the effect of alcohol must be true of all doses, large or small, although the effect of very minute doses may be imperceptible. Arsenic may be administered in doses so small as to produce no apparent ill effects; yet no one doubts that arsenic is a poison. . . . If a person dies of delirium tremens, it is not the last glass that kills him, but every dose or glass he has taken in his life has conduced to the result." — DR. REUBEN D. MUSSEY.

[1] See Note 6, page 361.

medical dispensatories, and by eminent medical writers too numerous to mention.

Certain poisons whose action is to deaden or paralyze the brain and nerves are called **narcotic poisons.** Alcohol is classed by the authorities among the narcotic poisons, because of its paralyzing effect on brain and nerve substance, of which we shall learn later.[1]

77. **The Alcoholic Appetite.**—A celebrated physician has said, "If a person eats bread three times a day for twenty years he is just as readily satisfied at the end of the time as he was at the beginning. Natural appetite, or hunger, is simply the demand for material to supply the growth or waste of tissue. Every substance capable of assimilation will satisfy that demand, and with that satisfaction ceases, for the time being, all relish for more."

This is not the case with alcohol. Its worst characteristic is its power to set up a perpetual craving for itself, which calls for repeated and increasing amounts,[2] This is called the **alcoholic appetite.** All natural appetites have natural limits. But the appetite for alcohol, created by the diseased conditions which it has itself produced, has no natural limit.

From the first glass of the boy just beginning to drink to the dram of the drunkard whose tissues are poisoned and inflamed by it, its nature is ever to excite a thirst for more. Whether it is drank in the form of wine, beer, cider, rum, or whiskey, its character is the same ; for the character of any substance depends upon its quality, not its quantity. The secret of the drunkard's craving for alcohol is in the nature of the drink, rather than in the weakness of the drinker.

[1] See Note 7, page 362.     [2] See Note 8, page 363.

There is a scientific connection between the first glass and the drunkard's fate. No one is safe who begins to take any liquor containing alcohol. Entire abstinence is the only safeguard against forming the alcoholic appetite, and the only cure for it when it is formed.

Because of the ease with which the alcoholic appetite is roused where it has been once formed, and the power of a little to form such an appetite, no liquor containing alcohol should ever be used as a flavoring for pies, pudding-sauces, jellies, or any other article of food.[1]

The habit of treating one's friends to beer, wine, or any other alcoholic drink, is simply asking them to poison themselves at our expense. Such treating is a mark of ignorance rather than an evidence of real courtesy or friendship, and must be so considered by one who understands the true nature of such substances.

Furnishing wines or liquors at parties, dinners, or other entertainments, or for guests or callers, is virtually offering poisonous drinks, and is never an act of true or intelligent hospitality or real kindness. It may be placing temptation too strong to be resisted in the way of an inherited or acquired appetite for alcohol.

78. **Cider.** — The fermented juice of apples, known as cider, contains alcohol varying in amount with the length of time it is allowed to ferment. When cider remains in air of the ordinary temperature, alcohol can usually be found in it in about six hours from the time it comes from the press, sometimes sooner. The power which alcohol has to create an appetite for itself makes a beverage containing it in any quantity an unsafe drink.

As cider grows older it is said to be growing hard; i.e.,

[1] See Note 9, page 365.

the amount of alcohol in it is increasing. Hard cider often contains ten per cent of alcohol. If a person begins, when the barrel of cider is first put in the cellar, to drink one or two glasses every day, and continues the same amount daily as the cider grows hard, he is every day getting more and more alcohol, which, besides benumbing his brain, hardening his heart, and spoiling his temper, has the power to create the craving that nothing but alcohol, in ever-increasing amounts, will satisfy.

Note this: Many drunkards have acquired the appetite for strong drink at the cider-barrel.

**79. Wine.** — Wine is made chiefly from the juice of grapes. It can be made from the juice of berries, currants, and other fruits. Drinks called home-made wines are frequently made from such fruits. But the ferments that produce alcohol are in the air and on the skins of the fruit ready to set up the pocess of fermentation as soon as their juices are pressed out. They cannot get inside the whole fruit to make alcohol, hence we are in no danger of being injured by alcohol when we take our fruit-juices fresh from the fruit itself.

Some wines called "light wines" contain a smaller amount of alcohol than others; but even the lightest wines contain alcohol enough to make them dangerous drinks. The power of alcohol to create an appetite for more does not change with the quantity.

The theory that the use of light wines will prevent the use of stronger drinks, and so diminish drunkenness, is disproved by the history of countries where such wines have come into general use. Where the alcohol in the light wines does not lead to a use of stronger drinks, enough more of the weaker drink is taken to produce an equal amount of drunkenness.   Almost whole communities, including men, women, and children, in some wine-making districts, will be in a continual state of drunkenness during a large part of the vinting season.[1]

80. **Beer.** — Certain of the fermented drinks are called "malt liquors."   These are beer, ale, and porter, which are made from barley and other grains.   The starch of the grain is turned to sugar by keeping the grain warm and moist until it is sprouted.   Heat is applied to kill the sprouts; and the grain, then called malt, is ground or mashed, and soaked in water.   To the sweet liquid thus obtained is added yeast, which is a kind of ferment, and hops.   The yeast sets up vinous fermentation, which changes the sugar of the wort to alcohol and carbonic acid gas.   The gas escapes in bubbles, producing a froth on the top of the fermenting-vat.   The alcohol does not escape: it remains in the beer, making it a poisonous drink.

**Beer** is responsible for many crimes.   It seems to have a benumbing effect upon the moral nature that prepares the drinker for wicked and cruel deeds and for deliberate crime.[2]   A copious beer-drinker often looks the very picture of health, and boasts of the healthfulness of his favorite beverage; but the testimony of physicians, surgeons, and life-insurance companies is

---

[1] See Note 10, page 365.

[2] " Beer and wine do not give strength for work, but, on the other hand, often make people dull, heavy, stupid, and unfit for work.   The most severe and continued work can be performed without them, and there are now some millions of people in this country who never taste them.   Happy will be the day when they are not drank by any, but particularly by the workingman, who finds it difficult to maintain his family.   Then will there be less quarrelling, poverty, and crime, and more food, clothing, and education."—DR. EDWARD SMITH *of London.*

that the beer-drinker, of all others, is most liable to swift and sudden death from some slight causes.[1]

The surgeon dreads him for a subject, for his blood is often in such a state that a slight cut may develop into a gangrenous wound that ends quickly in death. A slight cold brings on a fatal pneumonia in spite of the best physician's efforts. The president of a leading life-insurance company said once of this class of drinkers : —

"It was as if the system had been kept fair outside, while within it was eaten to a shell, and at the first touch of disease there was utter collapse : every fibre was poisoned and weak. And this, in its main features, varying, of course, in degree, has been my observation of beer-drinking everywhere. It is peculiarly deceptive at first, and it is thoroughly destructive at the last."

Various kinds of domestic drinks are sometimes prepared by adding yeast to sweet liquids in which roots or bitter herbs have been steeped. Those who make them are frequently not aware that they contain alcohol ; but the yeast acts on the sugar of the sweet liquid, just as it does on the sweet grain-juice, and changes it to alcohol and carbonic acid gas.

**81. Distilled Liquors.** — By heating a fermented liquid, the alcohol in it can be readily driven off in the form of vapor. This is condensed in a cool receiver, and the result is a new and stronger liquor. This process is called **distillation.** The alcohol of commerce is distilled mostly from whiskey.

Beverages thus distilled from liquids which contain alcohol are commonly known as "spirits" and "ardent (burning) spirits." Brandy is distilled from wine, and

[1] See Note 11, page 367.

rum from fermented molasses. Distilled liquors contain from forty to fifty per cent of alcohol, — the rest being water, flavored with various aromatics.

When the beer, wine, or cider drinker has acquired such a craving for alcohol that the amount in those weaker drinks no longer satisfies him, he resorts to these stronger liquors. These are still more harmful; and even if they do not bring the drinker swiftly to drunkenness, they so injure his body and mind that he never realizes the happiness, respect, and usefulness that he might otherwise have enjoyed.

In place of happiness he has misery; instead of respect, contempt; and instead of usefulness, he becomes a burden to others. He sees, when too late, how different his life might have been but for his first glass, that led on to others and their consequences.

**A Costly Vice.** — The annual report of the Commissioner of Internal Revenue for 1891, shows that the business of manufacturing intoxicating liquors has grown in the past year to a magnitude which it had never reached before. The spirits produced and deposited in distillery warehouses during the last fiscal year amounted to about one hundred and sixteen million gallons, and the quantity of spirits in the distillery warehouses at the end of the year was one hundred and thirteen million gallons, — the largest quantity ever known in these places. Outside of the warehouses there were, according to the figures of the Internal Revenue Bureau, about one hundred and fifty-three million gallons of spirituous liquor in the country. A total of two hundred and sixty-six million gallons of spirits, to be consumed chiefly by the people of the United States!

It would require a vivid imagination to conceive a tithe of the crime, misery, suffering, wretchedness, and death, that are included in that vast bulk of intoxicating liquor. It is estimated that the liquor consumed in the country costs the people one billion dollars a year.

# CHAPTER VII

## DIGESTION

**82. General Plan of Digestion.** — Food cannot serve the needs of the body until it has been converted into blood. In order to become blood, food must first be dissolved. This is done in the **alimentary canal.** This canal is a long tube, which runs through the body from the lips to the lower end of the spine.

Beginning at the mouth, this canal, continued as the *gullet*, passes down through an opening in the diaphragm, which forms the partition between the chest and abdomen. Below this it swells out into a large bag called the *stomach*. It then narrows again into the *small intestine*, which lies coiled upon itself in the cavity of the abdomen. It then swells out again into the *large intestine*, from which are discharged the useless parts of the food, together with certain other waste products.

This long digestive tube is lined throughout by a soft, moist, reddish kind of very thin skin, called mucous membrane. Layers of muscles surround the tube, which strengthen it, and help push along the food. In and around this tube are hollow organs, called glands.[1] They pour out certain fluids, which

[1] Glands are curious organs, of various shapes and sizes, whose special work it is to take out of the blood something to be used again, or to rid the blood of something to be cast out of the body. Thus the salivary glands make saliva, or spittle, and the sweat glands make sweat. The liver, which weighs about five pounds, is a single gland, and secretes bile; while the glands in the intestines are so very small that they cannot be seen by the naked eye.

alter the food chemically, and so fit it for absorption
into the blood.

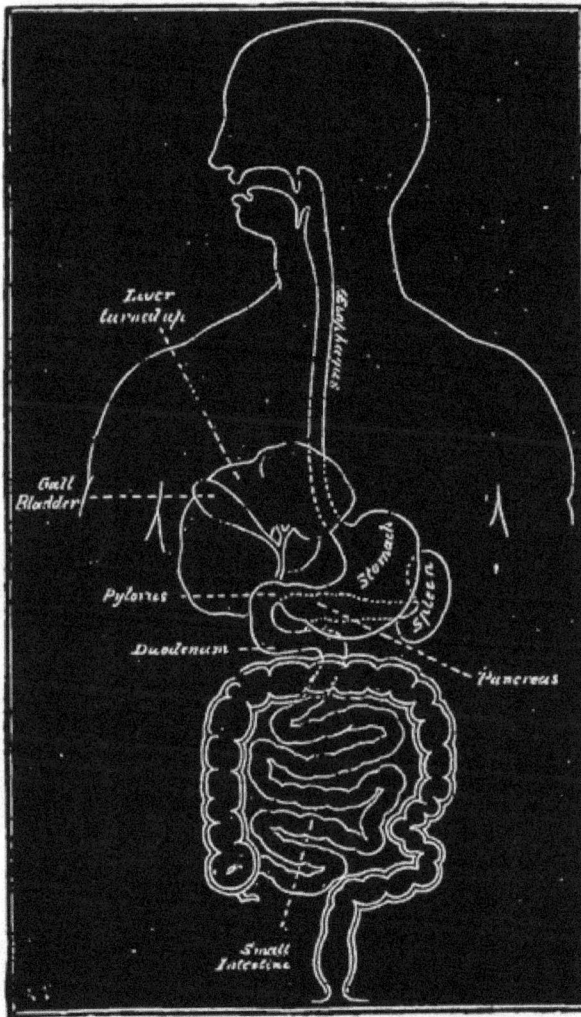

FIG. 35.— Blackboard Diagram of the Digestive Canal.

The process by which the food is made fit to mix
with the blood is called **digestion.**

83. **What takes Place in the Mouth. Chewing of**

the Food. — The **mouth** is a hollow box, with a movable floor formed by the tongue and lower jaw. At the top is the hard palate ; in front, the lips and the teeth ; and at the sides, the cheeks and the teeth. The moment food is taken into the mouth, it is rolled over by the tongue, and crushed and ground into very small pieces by the teeth.

This process is called **chewing,** or **mastication.**

84. **The Teeth.** — The **teeth** are small, hard, white, bone-like bodies fixed in the jaws, pressing against each

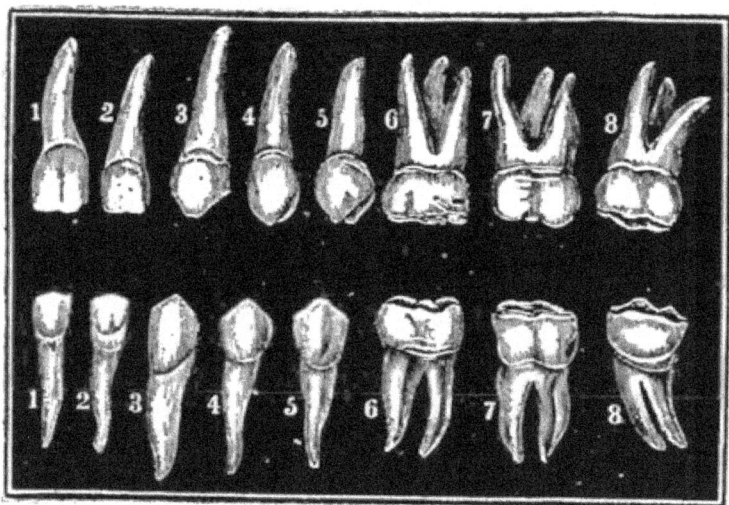

FIG. 36. — THE ADULT TEETH. — 1, 2, The cutting teeth (incisors). 3, Canine (cuspid). 4, 5, Small grinders (bicuspids). 6, 7, 8, Grinders (molars).

other as the jaws work, so as to cut and grind the food. They are fastened into the jaws by roots, which sink into the bony sockets somewhat in the same way as a nail is held in a piece of wood.

Teeth are made of three things, — *dentine, cement,* and *enamel.* Dentine, the familiar ivory of commerce, is a

bone-like substance, which forms the inside and body of
the tooth. Outside of this dentine is a layer of cement ; but, when the tooth appears above the jaw, the
enamel takes its place. It is a hard, shining kind of
polish, looks like ivory, and gives a strong protection
for the exposed part of the tooth, called the " crown."
Inside of each tooth is a space which holds a delicate
substance called the pulp, well supplied with nerves
and blood-vessels, which enter at the root of the tooth.

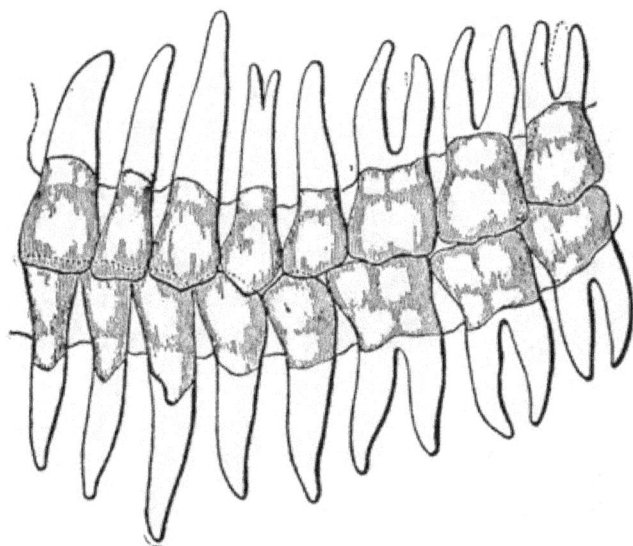

FIG. 37.—Diagram showing the apposition of the teeth.

A child when born has no teeth : afterwards, however, it has two sets. The first set, or milk-teeth,
twenty in number, are shed in childhood. The second
set, or permanent teeth, thirty-two in number, gradually take their place. The teeth are arranged in the
same way, and number the same, in each jaw, and in
each half of each jaw.

Thus, beginning at the centre of the jaw, there are eight *incisors*, or cutting-teeth, or two on each side. They have sharp, chisel-like, cutting-edges, which cut up the food. These teeth are largely developed in gnawing-animals, such as rabbits, squirrels, rats, and the beaver. Then come the *canine*, or dog-teeth, two in each jaw, so called because they are strongly developed in dogs, cats, tigers, and other flesh-eating animals.

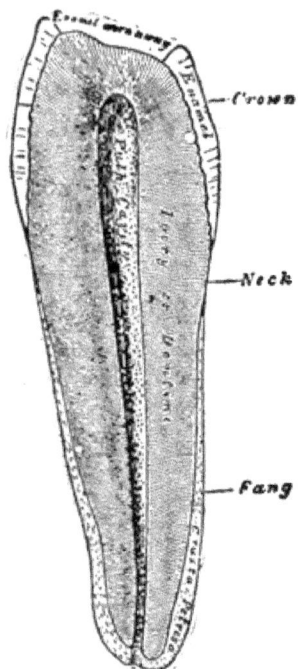

FIG. 38. — Vertical Section of a Bicuspid Tooth (magnified).

Next in order come the two-pointed teeth, called the *bicuspids*, four in each jaw. They are so called because they have two points or cusps, for grinding the food. After these come the largest and strongest teeth in the head, which do the hardest work. They are called *molars*, or "grinders." There are six of them in each jaw. They have broad crowns, with four or five cusps or ridges for grinding. The last molars are commonly called the "wisdom-teeth;" because they do not usually appear before the age of twenty, or the "age of wisdom."

**85. Mixing the Food with Saliva.** — While the food is being chewed, it is moistened by a liquid called the saliva, or spittle, which flows into the mouth from six little glands. They are known as the salivary glands. There are three of them on each side, in the following positions : The *parotid gland*, below and in front of each ear ;[1] the *submaxillary*, under the lower jaw ; and the *sublingual*, under the tongue, in the floor of the mouth. Each gland opens into the mouth by a little duct. They have a certain resemblance to a bunch of grapes, with a tube for a stalk.

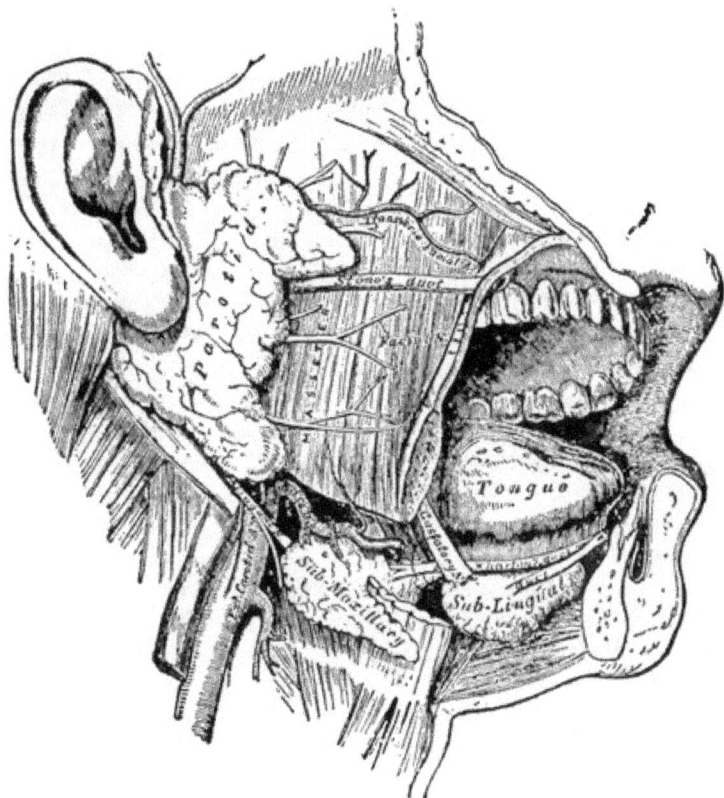

FIG. 39. — Salivary glands, with their ducts, on the right side.

Saliva is rapidly poured into the mouth while the grinding of food is going on. Three or four pints of saliva

[1] This gland, especially in childhood, sometimes becomes inflamed and swollen. This disease is familiarly known as the "mumps."

are secreted daily. Sometimes these glands are busily at work, even before we actually taste food. Touch
the tongue with the finger, and the saliva will flow : it
also pours into the mouth, even at the sight, smell, or
thought of food. This is what is commonly called
" making the mouth water."

**What is meant by Secretion.** — It is necessary to explain at this
point the exact meaning of secretion and excretion. The word
**secretion** comes from a Latin word which means to "sift or separate : " **excretion** comes from the Latin, and means " to sift out from."
Both words are used to express the sifting of some substance from the
blood.

A secretion is something taken from the blood to be used again in the
body for some special purpose ; while an excretion is waste matter, and
is thrown out of the body entirely. Thus, the salivary glands secrete
the saliva, and the liver secretes the bile. Sweat is an excretion, thrown
off from the body by the sweat-glands.

Those parts of the body which are engaged in secretion and excretion
are called *glands*. There are a great number of glands in the body,
varying in size, from the liver, which weighs about five pounds, to the
small glands in the intestines, which are invisible to the naked eye. All
glands, from the simplest to the most complex, have the same foundation
structure. They are built, so to speak, of identical material. They consist essentially of a fine skin, or membrane, which is covered on one side
with a layer of exceedingly minute bags called epithelial cells ; while it
rests upon capillaries on the other side. These cells have the power of
drawing the peculiar secretion of the gland out of the blood that is circulating through the capillaries. The moist, tender skin, or lining, on
the inside of our lips, is a familiar example of this secreting-membrane.

A gland may be merely a pouch in the basement membrane ; or the
membrane may be elaborately twisted, branched, made up like bunches
of grapes, and so on, into all kinds of complex glands. Why and
how the cells in the liver make bile, and not gastric juice ; those in the
stomach make gastric juice, and not sweat ; those in the skin make
sweat, and not saliva ; and so on, — we do not know, and we can hardly
imagine.

**Excretion** will be described in full in chapter xiii.

The saliva not only softens and mixes with the food, and keeps the mouth moist, but it also has the power of acting upon all starchy matters in the food, and changing them into sugar. These starchy matters, as we have seen, before they are mixed with saliva, are insoluble ; but, when changed into sugar, they are quickly dissolved in the stomach, and taken up by the blood.[1]

**86. How Food is swallowed.** — The food is now ready to be swallowed. The soft, moist mass is carried backwards by the tongue and the muscles of the mouth into the funnel-shaped top of the gullet, called the **pharynx.** The soft palate pushes up and backwards, so as to prevent the food from passing into the nose.[2]

Now, besides the opening from the pharynx into the stomach, there is also one into the windpipe. To prevent the food from getting into this opening, and choking us, the top of the windpipe is protected by a little lid, a kind of trap-door, called the *epiglottis.*

When we swallow, the tongue, raised and pushed backwards, shuts this little lid ; and thus a bridge is

[1] We can easily prove that the saliva changes starch into sugar: hold some boiled starch — or a little arrowroot, which is almost pure starch — in the mouth for a few minutes. We find that it gradually loses its thick, pasty, nature, and becomes thin and watery, while at the same time it acquires a sweet taste. The fact is, the saliva has acted on the starch, and changed it into sugar in order to make it soluble. This shows the necessity of properly chewing our food. If we bolt our food without sufficiently chewing it, the saliva does not act properly upon it, and a great amount of extra work is thrown on the intestines.

[2] During and after a severe attack of diphtheria, scarlet-fever, etc., in which diseases the parts of the throat are apt to be partially paralyzed, the soft palate is not able to shut off this passage into the nose: as a result, milk and other food often come up through the nose.

made, over which the food passes downwards into and through the gullet, and thence into the stomach. Sometimes, however, a morsel of food "goes the wrong way," — that is, is drawn into the opening of the wind-

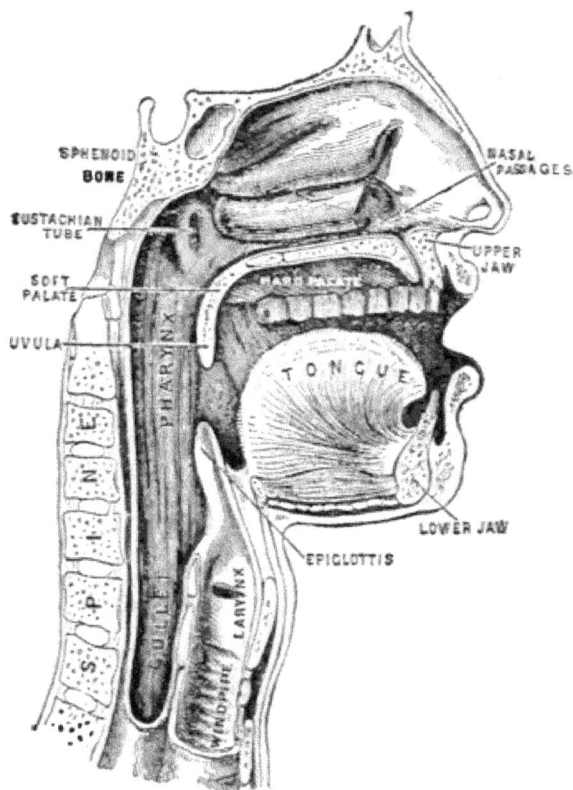

Fig. 40. — Section showing Passages to the Gullet and Windpipe.

pipe, or down into the air-tubes, — and then violent coughing follows : by this means it may be brought up again. If the substance is hard and large, like a boot-button, orange-seed, or peanut, a person, especially a child, may be choked to death.

The **gullet,** or food-pipe, is a tube about nine inches long, hanging loosely behind the windpipe. Its thick walls are provided with hoop-like muscles, which contract with a worm-like motion, well seen when a horse is drinking water, and so push the food along towards the stomach. The pellet of food is passed along from above downwards by these muscles[1] in some such way as we would push a ring along inside of a rubber tube.

**87. Digestion in the Stomach.** — The food, a moistened, partly-digested mass, has now reached the stomach.

The **stomach** is a pear-shaped bag, or pouch, capable of holding about two quarts of liquid. When full, it is some twelve inches long, and four inches broad. It is placed across the upper part of the abdomen, directly under the diaphragm, the larger end to the left side.

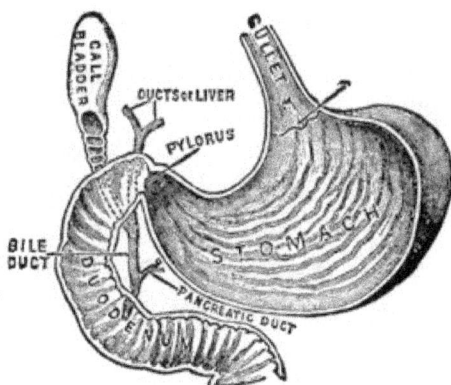

FIG. 41. — Section of Stomach.

[1] It is important to remember, that, in swallowing, the food and drink do not simply *fall* down the gullet. Their passage is controlled by the muscles in such a way that they grip successive portions swallowed, and pass them along. Even in swallowing a pill, there is the same process. The smaller the pill, the greater the difficulty in swallowing; because the muscles have more trouble to get the necessary grasp on it. Some of us have seen an acrobat or juggler stand on his head, and drink a glass of water, and even eat in this position. We see the same thing every day when we notice a horse or cow drinking from a pail of water on the ground.

When the food is pressed backwards by the tongue into the top of the gullet, it is no longer under the control of the will: it is impossible to recall the mass, and it is necessarily carried into the stomach.

The stomach has two openings. The opening, or ring, through which the food enters, and where the gullet ends, is called the *cardiac* orifice. The other, at the right end, where the intestines begin, and that by which food leaves the stomach, is called the *pylorus*, or gate-keeper.

The walls of the stomach consist of three coats : (1) A tough outer coat of fibrous tissue, which protects and strengthens the organ ; (2) a coat of smooth, or involuntary, muscular fibre, which rolls the food about in the process of digestion, with that worm-like motion which we noticed in describing the gullet ; (3) a third or inner coat, in which the blood-vessels and nerves are spread out.

This inner or mucous coat is next to the food, and has its surface honeycombed with millions of little pits. We have all seen this in tripe. In the floor of each of these tiny pits a number of tubes open. These tubes, during the process of digestion, are constantly pouring into the stomach a peculiar fluid, called the gastric juice.

The **gastric juice** is a clear, almost colorless, fluid, with a sharp acid taste. It contains a peculiar substance called *pepsin*, and an acid, both of which are necessary to the digestion of food in the stomach. The amount of gastric juice has been variously estimated, — all the way from five to fourteen pounds daily. There is comparatively little in the stomach at any one time. There is no loss, for it is rapidly re-absorbed by the blood.

Now, the moment the food reaches the stomach, the gastric juice begins to flow, the muscles begin to contract, and the whole organ takes on a churning motion.

The food is rolled over and over, and thoroughly mixed with the gastric juice. Two rings, one at the entrance and the other at the outlet, keep the food in the stomach while it is being churned about and digested.

The gastric juice has scarcely any effect on the starchy matters, but acts on the albumens, such as are contained in meat, cheese, eggs, etc., to make a complete solution.

FIG. 42. — The Stomach, showing its outer layers of muscular fibres.

Some of these dissolved albumens, together with some of the starchy matters dissolved by the saliva, are at once sucked into the blood by the blood-vessels in the walls of the stomach. The remainder of the mass, a thick, pulpy, soup-like substance, resembling gruel, is called **chyme**, meaning juice, with an acid odor and taste.

Chyme contains more or less albuminoid matters, together with the starchy elements which have not been changed into sugar, and all the fat. The fat may be seen floating in large drops on the surface of the chyme. The ring at the outlet, or pylorus, now relaxes ; and the partially digested food is forced in little jets into the duodenum, or first part of the small intestines.

**88. Digestion in the Intestines.** — The **intestines** consist of a long tube, or canal, which fills the greater part of the abdomen. They are divided into the small intestines, about twenty-five or thirty feet long, and the large intestines, five or six feet in length. To get all this into the small space it has to occupy, it is doubled upon itself many times. The first portion of the small intestines,[1] which is directly continued from the stomach, is called the *duodenum*, because its length is about equal to the breadth of twelve fingers.

Let us now see what goes on in the duodenum. Two pipes, or ducts, unite, and enter it. One comes from the gall-bladder, and brings the *bile;* the other from the pancreas, and brings the *pancreatic juice.* These two tubes unite, and enter the duodenum at the same place.

The **liver** is the large reddish-brown organ situated just under the diaphragm, and on the right and upper side of the abdomen. It is the largest organ in the body, and weighs about fifty ounces. The liver secretes

---

[1] The small intestine includes three parts, — *duodenum, jejunum,* and *ileum.* The large intestine includes the *colon, cæcum,* and *rectum.*

from the blood a greenish fluid called the bile. It is stored up in a kind of little pear-shaped bag attached to the liver itself, and called the gall-bladder.

The **bile** is a thin, greenish-yellow, bitter fluid. Sometimes it is olive-brown in color; but, when acted upon by the gastric juice, it takes on a distinctly yellow or greenish hue; hence the appearance of vomited bile.

The chief use of the bile is to digest the fatty parts of the food, upon which the gastric juice does not act. The bile also aids in separating the nutritious from the useless parts of the food.[1]

The **pancreatic juice** is very much like saliva. It is secreted by a long, narrow, flattened gland called the *pancreas*, or sweet-bread, which lies deep in the cavity of the abdomen, just behind the stomach. It is often said to resemble a dog's tongue in shape.

The pancreatic juice finishes the work which the saliva began. It acts chiefly upon the starchy matters of the food which have escaped the saliva, and changes them into sugar. At the same time, however, it follows up the work of the gastric juice, and dissolves any albumens which have not been dissolved in the stomach.

The inner surface of the intestines also pours in a liquid called the **intestinal juice;** which, like the saliva,

---

[1] The bile is alkaline; its chief ingredients are two salts of soda. We have tried, perhaps, to clean a bottle which has held oil. We put in some soda and warm water, and then shake the bottle.. The soda breaks up the oil into tiny drops, which float round the creamy mixture. This is exactly what the bile does with the fatty parts of the food. In other words, it makes an "emulsion." Cod-liver oil is often mixed with other things to disguise the fishy taste. This creamy mixture, known as an "emulsion of cod-liver oil," is a familiar sight in the drug-stores.

acts upon the starchy matters.   At the same time, however, this juice acts as a solvent on any albuminoid
matters or fats still undissolved.   In fact, this intestinal
juice has the properties of saliva, gastric juice, bile, and
pancreatic juice, all in one.

Like a prudent housekeeper, Nature intends that this
last digestive fluid in the digestive tube shall act upon
all matters in the intestines still left undissolved by the
other juices, so that no part of the food may be wasted.
The chyme thus acted upon by these different fluids
has the appearance of a thick cream, and is called
chyle.

The muscles of the intestines, like those of the
stomach, are almost continually at work.   The whole
intestinal tube is constantly moving in a manner somewhat resembling the motions of a worm; and, all this
time, it is squeezing and forcing forward its contents.
In this way, a portion of the food is acted upon by
the several parts of the intestinal canal.

89. Absorption.  How the Blood feeds on the Food.
—The soup-like mass which left the stomach under the
name of chyme, has now been changed into a thick
cream called chyle.   Squeezed slowly along the intestines by the worm-like twistings of the muscular walls,
most of the nutritious parts of this milky fluid is sucked
up drop by drop by the lacteals, and poured into the
current of the blood.   The process by which the digested materials are taken into the blood is called absorption.

This is done chiefly by two sets of vessels, — first, by
the *lacteals*, or *lymphatics;* second, by the *blood-vessels*.

**90. Absorption by the Lacteals and Lymphatics.** —
The inner surface of the intestines is not smooth and
shiny, like the outside, but shaggy, or, rather, has a velvety appearance. This is because the inner lining is
crowded all over with millions of little tags, like very
small tongues, hanging down into the inside of the
intestines. These hair-like projections are called villi,
meaning "tufts of hair." They are tiny affairs, about
one-thirtieth of an inch long; and a five-cent piece
would cover five hundred of them. These give the
appearance of the pile
on velvet. We are, of
course, familiar with
this appearance in tripe.

In each one of these
villi is a network of the
finest blood-vessels, and
a tube, or canal, called a
lacteal, so called from
a Latin word meaning
"milky," and because

FIG. 43. — Two Villi (magnified).

it carries a white, milky fluid. Millions of these lacteals
dip down into the intestines, like little root-fibres,
and suck up from its creamy contents the fatty matters
of the chyle.

The **lacteals,** after passing through a number of
glands — like way-stations on a railroad — in the abdomen, unite into larger tubes, and finally open into a
little sac, or bag, in the loins, called the receptacle of
the chyle.

Leading upward from this is a tube called the **thoracic**

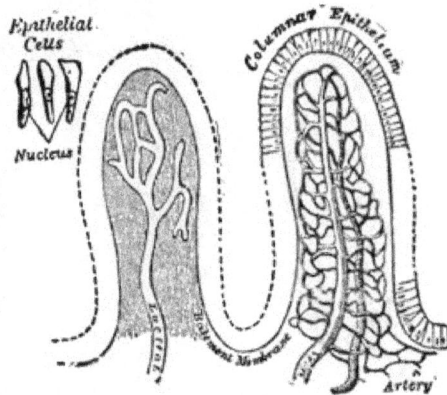

duct, which carries the fluid upwards, along the backbone, and pours it into a vein in the neck. The thoracic duct is about eighteen inches long, and about as
large as a goose-quill. It acts as a feeding-pipe, to
empty the nutritive matter absorbed from the food into

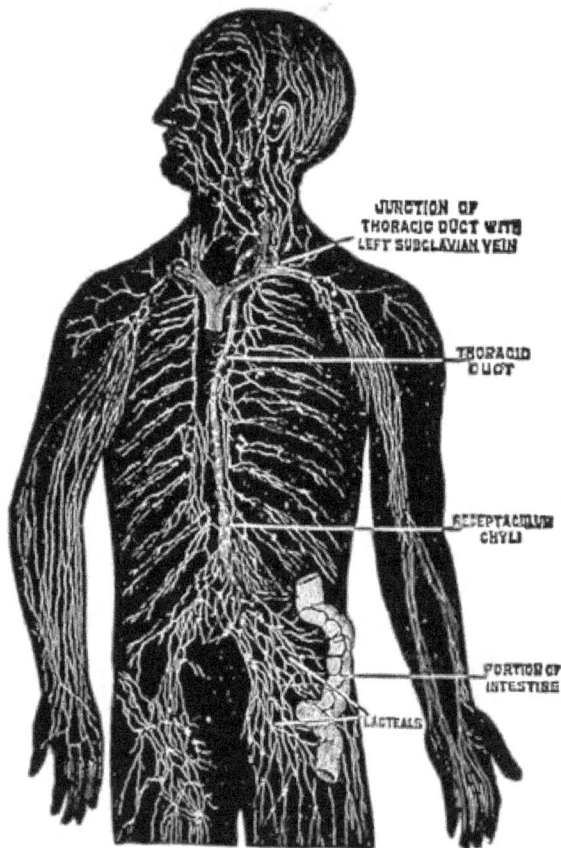

FIG. 44. — Lacteals ending in Thoracic Duct, which empties into a Vein in the Neck.

the blood-current. The lacteals are merely lymphatics,
which begin in the villi of the intestines.

In all parts of the body, except the brain, spinal cord,
eyeball, and tendons, we find thin-walled vessels busily

at work taking up, and making over anew, waste fluids or surplus materials derived from the blood and tissues generally. They seem to start out of the part in which they are found, like the rootlets of a plant in the soil. The tiny roots join together, and make larger roots. They carry a colorless fluid called *lymph*, very much like blood without the red corpuscles. The tubes in which this fluid flows are called **lymphatics.**

These tubes have rounded bodies at many points of their course, scattered like depots along a line of railroad, called *lymphatic glands.* They seem to be factories to make over the lymph in some way, and to fit it for renewing the blood. Most of the lymphatics at last unite with the lacteals, and empty their contents into the thoracic duct.

Thus, again, Nature, like a careful housekeeper, uses up even the waste matters of our bodies, in providing them with nourishment.

**91. Absorption by the Blood-vessels.** — Soluble substances can pass through the delicate walls of the blood-vessels. The inner lining of the digestive canal, especially the small intestines, is rich in blood-vessels.

---

**The Spleen.** — The spleen is a reddish, soft, and pulpy organ. It is about five inches in its longest diameter, and an inch thick, and situated near the left end of the stomach, just under the ninth, tenth, and eleventh ribs. The use of this organ is not known. It has probably something to do with the formation of blood-corpuscles. In certain diseases, like malarial fever, it may reach an enormous size. Two other glands, the *thymus* and *thyroid glands*, known when taken from the lower animals as the "breast sweet-bread" and "throat sweet-bread," are situated in the neck. They are regarded as appendages of the lymphatic system. Their use is not known.

Hence those parts of the food which have been made liquid, and capable of forming a soluble mixture with the blood, are absorbed directly by the small veins and capillaries ; although a small part finds its way into the lacteals.

The emulsified fats, as we have seen, are mostly taken up by the lacteals. Certain materials, dissolved by the gastric juice and ingested liquids, are taken up directly by the blood-vessels of the stomach. An enormous absorbing surface is provided in the small intestines, by the arrangement of the mucous membrane in folds, and by the countless numbers of villi, which are constantly absorbing nutritive materials through the myriads of their tiny blood-vessels.

It is important to remember that all the blood circulating in the digestive organs, and taking up the soluble nutritive matters, must pass through the liver before entering the general circulation ; and from it the liver selects and elaborates its secretion.

92. **The large Intestines.** — The large intestines begin at the lower extremity of the small intestines, and consist of three parts, the ascending, the transverse, and the descending *colon ;* the last of these being continued into the *rectum.* This digestive tube, which is about six feet long, first passes up the right side of the abdomen, then across under the liver and stomach, and, lastly, descends on the left side of the abdomen to the rectum.

As the food-mass passes along into the large intestines almost everything which is fit to be taken up has been absorbed. Some digestion may go on in them, but it is of little account.   They are really a kind of passage-way, or temporary storehouse, for undigested and waste matter, which should be cast out of the body as speedily as possible.

**93. How much to Eat.** — The quantity of food needed to keep the body in good health varies very much according to circumstances.   The general principles are these : Action is the law of every living being.   Waste attends action: hence, in the main, the supply of food must equal the waste of the body.   The greater the amount of exercise, the more food is called for to supply the waste.   During the time of growth, a still greater quantity is needed to build up new tissues ; hence growing children generally have a good appetite and a vigorous digestion.   The same holds good of persons getting well of some long and wasting sickness.

The quantity required also depends very much upon one's business.   Those who work hard and long, either with the body or mind, as teamsters, blacksmiths, farm-ers, doctors, and editors, need a goodly amount of nutritious food.   Those who work in-doors, as clerks, milliners, students, and book-keepers, can get along with a smaller quantity.   In cold weather, or in cold climates, a greater quantity of "fuel-food" is necessary than in warm weather or in a tropical climate.[1]

[1] "Travelers' accounts of the amount of food consumed by the natives of the frigid zone are almost incredible.

"Dr. Hayes, the Arctic explorer, states from his own observation, that the daily ration of the Esquimos is from twelve to fifteen pounds of meat, about one third of which is fat.   He once saw an Esquimo consume ten pounds of walrus flesh and blubber at a single meal, which however lasted several hours, with the thermometer 60° or 70° below zero.   Some members of his own party manifested a constant craving for fatty substances, and were in the habit of drinking the contents of the oil-kettle with evident relish." — *Flint's Physiology.*

An appetite for plain, simple, well-cooked food is a safe guide to follow.   Every person in good health and with moderate exercise should eat well : he should have a keen appetite for his food, and enjoy it.   Young, grow-ing, and vigorous persons should eat until appetite is fully satisfied, provided they have enough of exercise, both mental and bodily.

It is easy to know when we are eating too much.   An overworked stomach makes its condition known by a sense of fulness, uneasiness, drowsiness after meals, and sometimes a real distress.   If we keep on eating too much and too rich food, the complexion is apt to be muddy, the face more or less covered with blotches and pimples, the breath has an unpleasant odor, and the general look of the face is dull and unwholesome.

94. **What to Eat.** — Food should be both nutritious and digestible.   It is nutritious in proportion to its capacity to furnish suitable substances to be taken into the blood.   Foods are digestible just as they are acted upon by the digestive fluids.   Certain foods, as the vegetable albumens, are both nutritious and digestible. A man will grow strong and keep healthy on any of them.[1]

[1] The splendid races of Northern India live on barley, wheat, millet, and rice, as their staple food.   In Southern India, millions of people live on pease and rice.

There have been many generations of the hardiest men in the world in the north of England and Scotland, who lived on oat-meal and milk.   These men literally lived their lifetime without taking any other food except green fruits, vegetables, and the cereals.

The Roman gladiator's chief article of diet was barley; and the ancient sol-dier endured his long marches and severe fighting on the cereals boiled in water, when meat could not be had.

We can safely eat some animal food every day, yet it is well to remember that the vegetable albumens supply all that is needed for the nourishment of the body. A strong, hearty person may eat half a pound or so of meat daily ; yet he should take other foods, such as bread, oatmeal, beans, rice, and milk. These foods are all good, all cheap, all digestible, and all palatable.

Vegetable foods are less stimulating than animal. Hence they are more suitable for children. Beef, pork, ham, oysters, and rich pastry should be sparingly given to children. The plainest and simplest diet is the best. It is much better for a child to go to bed on a supper of oatmeal, baked apples, or mush and milk, than warm bread, cake, pie, and fried meat. Parents cannot look too sharply after the food they give their children. This is especially important at the present day, when all kinds of strange devices are resorted to, to furnish the public with fanciful and tempting foods.

Students must also look sharp to their diet. It is much better for them to begin a day's study with a breakfast of oatmeal, stale bread, a soft-boiled egg, and a glass of milk, than strong coffee, sausage, and hot biscuit.

95. **When to Eat.** — Three meals a day, from five to six hours apart, should be eaten. These should be arranged mainly according to one's occupation. The stomach, like other organs, does its work best when its tasks are done at regular periods. Hence, regularity in eating is of the utmost importance.

Eating out of meal-times should be strictly avoided, for it robs the stomach of its needed rest. Food, eaten when the body and mind are tired, is not well digested. Rest, even for a few minutes, should be taken before eating a full meal. It is a good plan to lie down, or sit quietly and read, fifteen minutes before eating.

Severe exercise and hard study just after a full meal, are very apt to check digestion. The reason is plain: after a full meal, the vital forces of the body are called upon to help the stomach digest its food. If they are forced, in addition to this, to help the muscles or brain, digestion will be hindered, and a feeling of dulness and heaviness follows. This, in time, often results in the common form of ill digestion called "dyspepsia."

We should not eat for at least three hours before going to bed. When we are asleep, all the vital forces are at a low ebb, and digestion is difficult, if not impossible.

We should make it a point not to omit a meal unless forced to do so. Children, and even grown-up people, often have the bad habit of going to school or to work in a hurry, without eating any breakfast. There is sure to be an "all-gone" feeling at the stomach before another meal-time.

The state of the mind has a great deal to do with digestion. Sudden fear or joy, or unexpected news, may take away the appetite at once. Hence, so far as we can, we should laugh and talk at our meals, and drive away all anxious thoughts and unpleasant topics of discussion. If hunger is a good sauce, so also is a jolly laugh.

96. **How to Eat**. — Eat slowly, and thoroughly chew the food. Do not take too much drink with the food.

Our teeth were made to chew our food, and the saliva to moisten it, and help along digestion. If the food is well chewed, the saliva and the gastric juice act more readily. It is not only bad manners to eat rapidly, but it is a violation of the simplest law of digestion.

If we take too much drink with our meals, the flow of the saliva is checked, and digestion is thus hindered. Rapid eating, with a great deal of drink to wash down the food, is almost sure to result in dyspepsia.

Do not take the food and drink too hot or too cold. Such substances as very hot bread and coffee often injure the enamel of the teeth, and are slowly digested in the stomach. If the food and drink are taken too cold, undue heat is taken from the stomach, and diges-

**Proper Care of the Bowels.** — Irregularity in eating, too much finely-bolted flour, and not enough fruit and vegetables, rich pastry, negligence or carelessness in attending to a regular daily evacuation of the bowels, lead to the very common and distressing trouble known as *constipation*, or costiveness. As we grow older, this trouble generally grows worse. Many persons resort to all kinds of patent pills and medicines, which, for the most part, act only for the time, and do not remove the cause of the ailment. Such remedies do a deal of harm after a time, and are almost sure to leave the bowels in a worse condition.

We must pay strict attention to the proper action of the bowels. The formation of a regular habit is of the utmost importance. The bowels can be trained to act at a certain time every day. Take great pains to eat coarse food, such as oatmeal, corn-bread, vegetables, stewed prunes, dates, figs, etc. Drink a glass of water just after getting out of bed in the morning. Rochelle or Seidlitz powders may be needed at times; they are safe, and will do no harm. Vigorous muscular exercise, especially kneading the muscles of the abdomen, is a valuable help.

If these things do no good, consult a physician before purgative medicines are used in a hap-hazard sort of way, simply because some friend urges their use.

tion delayed.  The natural temperature is about 100°
F.  If we drink freely of ice-water or cold well-water,
its temperature will fall about 30°; and it will take half
an hour or more for the stomach to regain its natural
heat.

Drinking freely of very cold water when the body is
heated is a dangerous practice, and, aside from its ill
effects on digestion, has occasionally resulted fatally.  It
is a poor plan to bolster a flagging appetite with highly-
spiced food and bitter drinks.  An undue amount
of pepper, mustard, horse-radish, pickles, fancy meat-
dressings, and highly-seasoned sauces, may stimulate
digestion for the time, but, used in excess, they soon
weaken it.

97.  The Proper Cooking of Food. — To prepare, cook,
and serve food well, is a fine art.  To voluntarily leave
this important work to raw and untrained servants, is
an evidence of false notions, and of a neglected or one-
sided education.

A knowledge of the principles and practice of cooking should form a part of every young person's education.  Many an unhappy home is due to the neglect of
the orderly and cleanly housekeeping and good cooking
that raises the housekeeper to the dignity of the home-
maker.  Many a man has been led into taking alcoholic
drinks to quiet a craving caused by his system being
imperfectly nourished.  He is really hungry, not because of any lack in the quantity of his food, but
because of its poor quality.

Many a constitution that might otherwise have been
sufficiently strong and healthful to meet the demands

of life, has been ruined ; and dyspepsia and other forms of indigestion have resulted from the use of fatty, fried, soggy, heavy, and otherwise unwholesome food due to poor cooking.

Every young woman who goes into a home of her own, from the office, shop, factory, or any position or occupation where she has had little or no practice in cleanly, orderly housekeeping, including good cooking, should take a special course of instruction in these matters so important to the health and welfare of a household.

98. **Care of the Teeth.**—The care of the teeth is an important matter. It is our duty to take the very best care of them, and to keep them as long as possible. Teeth are prone to decay. We may inherit poor and soft teeth : our ways of living may make bad teeth worse.

If an ounce of prevention is ever worth a pound of cure, it is in keeping the teeth in good order. They should be thoroughly cleansed night and morning with a soft brush and warm water. Castile soap, and some simple tooth-powder with no grit in it, may be occasionally used. Dentists say that we should always cleanse the teeth before going to bed. The brush should be used on the inner or back side of the teeth, as well as on the front side.

The enamel once broken or destroyed is never renewed. The tooth is left to decay, slowly but surely : hence we must be on our guard against certain things which may injure the enamel. Picking the teeth with pins and needles is hurtful. We should never crack nuts, crush hard candy, or bite off stout thread with the teeth. Metallic tooth-brushes, gritty and cheap tooth-powders, and hot food and drink, often injure the enamel.

We should never use any of the tooth-powders and washes, especially advertised and secret preparations, warranted to harden the gums, and whiten the teeth. Dirty and decayed teeth are a frequent cause of an offensive breath and a foul stomach.

We should exercise the greatest care in saving the teeth. The last resort of all is to lose a tooth by having it extracted. The skilled dentist will save almost anything in the form of a tooth.

99. **Effect of Alcohol on the Stomach-Digestion.** — Alcoholic liquors act as a mild or strong irritant of the stomach, just as they are taken, raw or diluted. Their habitual use leads to most distressing forms of stomach disease.[1] If we could look into the stomach, as Dr. Beaumont looked into the stomach of Alexis St. Martin,[2] just after taking a drink of raw spirit, we should find that the inner surface would be bright red when the alcohol touched it, and far more so than after taking ordinary food. Alcohol irritates the lining of the

---

[1] " Many cases of dyspepsia are due to alcohol, solely and wholly." — J. MIL-NER FOTHERGILL, M. D.

" Nothing with such certainty impairs the appetite and the digestive power as the continued use of strong alcoholic liquids." — PAVY *on* " *Food.*"

[2] The process of gastric digestion was studied many years ago by Dr. Beaumont and others, in the remarkable case of Alexis St. Martin, a French-Canadian, who met with a gun-shot wound which left a permanent opening into his stomach, guarded by a little valve of mucous membrane. Through this opening the lining of the stomach could be seen, the temperature ascertained, and numerous experiments made as to the digestibility of various kinds of food.

stomach, and dilates the tiny blood-vessels, just as brandy dropped into the eye would make it look red and watery.

Alcohol, like any other irritant, also causes the gastric juice to flow, just as the eye, when injured, becomes flooded with tears. The power of the gastric juice to dissolve food is at the same time either greatly diminished or entirely destroyed. For, whenever alcohol comes in contact with gastric juice, the pepsin, without which it cannot digest food, is precipitated. This hinders digestion until the alcohol is sufficiently diluted with water drawn from the stomach to prevent further precipitation of the pepsin.[1] Thus an extra amount of useless work is thrown upon the glands and makes them less active.

If this unnatural excitement of the glands of the stomach is kept up for some time, they become weakened, and the gastric juice is diminished in quantity and made poorer in quality. The result is alcoholic dyspepsia.

It has even been found by drawing off the contents of the stomach with a siphon, during various stages of digestion when alcohol has been taken, that it entirely suspends the transformation of food while it remains in the stomach, and that only after the alcohol leaves the stomach does digestion go on at a fair rate.

In the case of St. Martin it was observed by looking directly into his stomach that when a wholesome dinner was digesting in good order, a glass of gin arrested the process, which was not continued

1 See Note 12, page 368.

until after the alcohol had passed out of the stomach.

Now, when this alcoholic irritant is poured into the stomach for days, weeks, and even for years, it is no wonder that the stomach becomes altered in its structure.[1] There is a chronic inflammation of the inner coats: the walls of the stomach become thicker and harder, and traces of ulcerations are often found. Because the stomach is unable to digest food properly, many other organs of the body suffer as a result.

**100. Effect of Alcohol on the Liver.** — The liver is peculiarly liable to diseases due to alcoholic liquors.[2] When alcohol is taken up by the blood-vessels of the stomach, it is carried directly to the liver, and filtered through this great organ before it reaches the heart. This, as we have been told, is a part of the portal circulation. Hence the poisonous effects of alcohol are strongly marked in the liver, especially among hard drinkers, and in hot climates.

The blood-vessels of the liver are overworked, and the capillaries engorged with blood. This causes, first,

[1] "The structural alterations induced by the habitual use of alcohol, and the action of this agent on the pepsin, seriously impair the digestive power. Hence it is, that those who are habitual consumers of alcoholic fluids, suffer from disorders of digestion. Heartburn, water-brash, acid stomach, and a peculiar retching in the morning, are produced." — ROBERT BARTHOLOW, M.D.

"Alcohol in any appreciable quantity diminishes the solvent power of the gastric fluid so as to interfere with the process of digestion instead of aiding it.". — W. B. CARPENTER, M.D.

"If the amount of alcohol be increased, or the repetition become frequent, some part of it undergoes acid fermentation in the stomach, and acid eructations or vomitings occur. With these phenomena is associated catarrh of the stomach and liver with their characteristic symtoms — loss of appetite, feeble digestion, sallowness, mental depression, and headache." — DR. JAMES C. WILSON.

[2] See Note 13, page 369.

an enlargement of the liver, and then a shrivelling of the substance of the organ, together with a rough and "bunchy" surface. This, in medical language, is called *cirrhosis*,[1] meaning tawny, or orange-colored, but in common phrase it gets the name of "hob-nail" or "gin-drinker's" liver, from its appearance.[2]

Again, fatty decay (called fatty degeneration), due to alcohol, may take place in the liver: this makes the organ of great size, sometimes weighing from fifteen to twenty pounds ; and one case is related of a drunkard's liver that weighed fifty pounds.

**101. The Effect of Tobacco on Digestion.** — The use of tobacco, either in smoking or chewing, causes the glands of the mouth to send out large amounts of saliva ; this, in time, weakens them and causes dryness of the throat. It also interferes with digestion.[3] Very frequently smoking leads to indigestion, which can only be cured by abandoning the tobacco.[4]

---

1 "Cirrhosis of the liver is notoriously frequent among drunkards, and is in fact almost, though not absolutely, confined to them."–ROBERT T. EDES, M.D.

2 " Alcohol acts on the liver by producing enlargement of that organ, and a fatty deposit, or "hob-nailed" liver mentioned by the English writers."— W. B. CARPENTER, M.D.

3 " Tobacco impairs digestion, poisons the blood, depresses the vital powers, causes the limbs to tremble, and weakens and otherwise disorders the heart. Physicians meet with thousands of cases of *dyspepsia* connected with the use of tobacco in some one of its forms." — DR. MUSSEY.

4 See Note 14, page 371.

## CHAPTER VIII

### THE BLOOD AND ITS CIRCULATION

102. **Uses of the Blood.** — Every child knows, that if he cuts or scratches his finger, or even pricks it with a pin or needle, blood flows. What is true of the finger, is also true of every other part of the body, except the hair and nails.

The tiny blood-vessels go everywhere, through the muscles and nerves, over and within the brain, through every particle of every bone, — the blood flows everywhere. Every little bit of bodily tissue is bathed with the blood. The blood makes everything common as it flies from spot to spot. Wherever it goes, it has something to bring, and something to carry away. The blood lives on the food we eat, and the tissues live on the nutritive material from the blood.

The blood, in its ceaseless round, not only brings new material for repair, but it is also a kind of sewer-stream that drains off waste matters, and carries them to organs whose duty it is to cast them out of the body.

103. **How Blood is made up.** — Blood, as it is drawn from the body, is a red, somewhat sticky, fluid, thicker than water, and apparently all of one substance. It is not so simple a fluid as it looks. If we let fall a small drop of freshly drawn blood upon a piece of glass, and examine it under a microscope, we shall see that this fluid is not all of one substance. It consists of an almost color-

less fluid called the **plasma,** and an enormous multitude of little bodies called **corpuscles,** floating in the liquid. These corpuscles are of two kinds, **red** and **white.**[1]

104. **Blood Corpuscles.** — The red color of the blood is due to millions of little **red corpuscles** which float about in it. When carefully measured they are found to be about the $\frac{1}{3200}$ of an inch in diameter. The white corpuscles are larger, being about the $\frac{1}{2500}$ of an inch across.

FIG. 45. — Human Blood-globules; *a*, seen from the surface; *b*, seen from the side; *c*, united in rouleaux; *d*, rendered spherical by water; *e*, decolorized by the same; *f*, Blood-globules shrunk by evaporation.

The red corpuscles are so closely packed, that the blood, as a whole, looks uniformly red. In the same way, a clear white glass bottle, filled with red beads and water, would look uniformly red at a short distance. Imagine a small brook all alive with little red fishes, and it would give some idea how the red corpuscles make the blood appear red. The red corpuscles are by far the more numerous, there being about five hundred of them to one of the white corpuscles.

In shape the red corpuscles are flattened circular disks, resembling somewhat pieces of coin. They have something of the shape of an India-rubber air-cushion when blown up with air. They are not hard and solid,

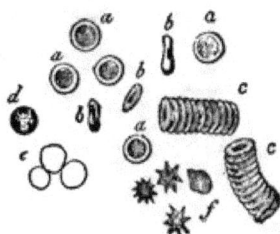

---

[1] The total quantity of blood in the body is about one-thirteenth of the weight of the body. Hence a man of average size has from ten to twelve pounds of blood. Of this amount, water makes up nearly four-fifths: only about one-fifth consists of dry solids. The solids consist mainly of the corpuscles, the albumen of the serum, and fibrine.

The blood also contains a small quantity of minerals and salts. The most important are iron, potash, soda, phosphorus, and sulphur. These are all essential to healthy blood.

but resemble tiny pieces of red jelly rather than anything else.

In size they are so very small, that, if we had fingers
delicate enough to handle them, we could pack away
some fifty thousand of them on the head of a pin. It
is said that some five million of them will float round
in a single drop of blood. Under the microscope, the
flat sides of these disks stick to one another in rolls,
piled together like so many gold dollars.

FIG. 46. — a, White Corpuscles of human blood; d, Red Corpuscles (high power).

The red corpuscles absorb oxygen in the lungs, and
carry it for distribution to the various tissues of the
body. Night or day, whether we are asleep or awake,
millions of these tiny oxygen-carriers are as busy as
bees. The blood has been beautifully called "the river
of life."[1] This is especially true of the red corpuscles.
They may well be compared to a countless fleet of little
boats which are constantly floating along "the river of
life" in our bodies.

[1] So dependent is all life on this fluid, that should the blood fail, for an
instant, to reach the brain, all consciousness would at once cease; and if for a
few seconds, life itself would cease.

The **white corpuscles** are slightly larger than the red, and are not flattened. As we watch them by means of the microscope, rolling and tumbling about, we see that at one time they are round like a base-ball, and of such a size that about twenty-five hundred of them would just reach one inch. Shortly afterwards, however, they change this form, and become pear-shaped, three-sided, and so on, in endless variety. The function of the white corpuscles is not certainly known.

**105. Coagulation, or Clotting.** — If a basin of fresh blood is allowed to stand for a short time, it will separate into two parts : one, a sticky, jelly-like mass, called the **clot,** settles to the bottom; the other, a straw-colored, watery fluid, called the **serum,** at the top. This change of the blood, after it is drawn from the body, into a jelly-like, semi-solid mass, is called clotting, or *coagulation*, of the blood.

The watery part, or serum as it is called, is blood out of which the corpuscles have been strained by the process of clotting. It is largely made up of water, with albumen dissolved in it. If we try to boil serum, we find we cannot do it. Before it boils, it "sets" into a stiff, solid mass, just like the white of a hard-boiled egg.

The clot consists mainly of two substances, — the corpuscles, and a network of white, tough, fibrous threads, called *fibrine*. It is owing to the presence of fibrine that the blood clots. The corpuscles get snarled into the meshes of the fibrine, and thus the clot is formed. Fibrine may be seen by whipping fresh blood with twigs, to which it will stick in fine threads.  The same thing can be done by beating up a bowl of fresh blood with an egg-beater, just as eggs are beaten for cake.

The power of coagulation is of the most vital impor-tance.  When a person receives a severe wound, he would bleed to death unless clotting set in.  Nature in this way plugs up the wound with clots of blood, and prevents excessive bleeding.

**106. General Plan of Circulation.** — If all the tissues and organs of the body stand in such constant need of blood, there must needs be some special machinery to furnish them with this vital fluid.  We shall find this to be the fact.  Let us begin to study this apparatus.

In the first place, there is in the chest a powerful forcing-pump, called the **heart,** from which pipes are distributed to all parts of the body.

One set of pipes, called **arteries,** carries the blood from the heart.

Another set of pipes, called **veins,** brings it back to the heart.

The manner in which blood is made to flow through vessels of the body may be compared to the way in which water is supplied to a city.  The heart is the pumping-engine which forces the blood into the main pipes for the supply of the several districts.

And as through the city the great water-mains branch and sub-divide into smaller pipes for the supply of districts, streets, houses, and rooms, so in the body the blood-vessels divide over and over again, to furnish a supply of blood to the smallest organs and the most minute parts of our bodily frame.  The parallel ends here. The water supplied to the city does not return to the pumping-station, whereas the blood returns to the heart.

When the blood has been pumped through every part of the body, and has given to it its nutritive supply, it receives from the tissues certain waste matters, the result of wear and tear. Hence the blood is no longer fit for nourishment. It is now more like a kind of sewer-stream laden with waste matters. These, in due time, are brought to certain organs, as the lungs, the skin, and the kidneys, and cast out of the body.

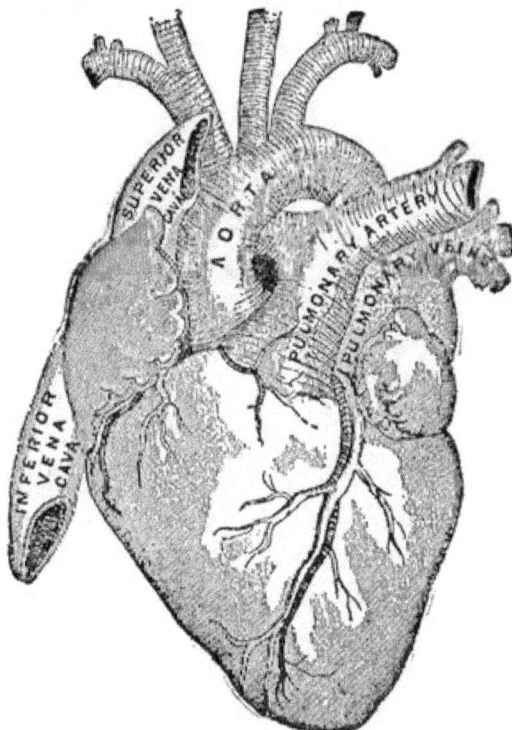

Fig. 47. — The Heart and its Large Blood-Vessels.

**107. The Heart.** — The heart is a hollow, muscular organ, somewhat like a pear in shape. It is hung almost in the centre of the chest, above the diaphragm, partly overlapped by the lungs, and opposite the breast-bone. It is about the size of one's closed fist, and is composed of involuntary muscular fibre.

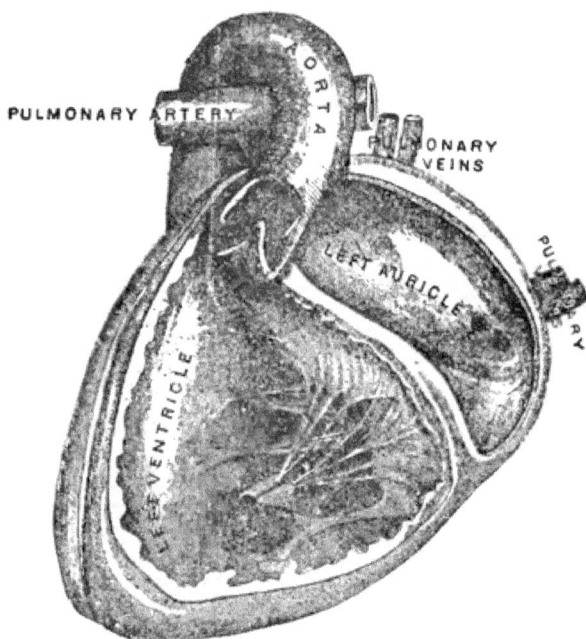

FIG. 48.— Left Section of Heart.

The heart is of a rounded, conical shape (some compare it to the shape of a cocoanut), and placed with the broad part uppermost, and the point slanting downwards and to the left, where it may be felt beating, between the fifth and sixth ribs.

The heart is a double organ, with a partition-wall running down the centre from top to bottom, which separates the right side from the left. Each of these sides has two hollow chambers, or cavities, — an upper" one called an auricle, from its fancied resemblance to an ear; the other and lower one called a ventricle. Hence there are two upper chambers called *auricles*, and two lower chambers called *ventricles*.

The heart is a muscle: hence it can contract. When each of its chambers contracts, blood is forced to flow into the next chamber or a blood-vessel, as the case may be. The walls of the ventricles are stouter and stronger than those of the auricles, and those of the left ventricle are much stouter than those of the right ventricle. The right auricle opens into the right ventricle, and the left auricle into the left ventricle; but there is no connection between one side of the heart and the other.

108. **Valves of the Heart.** — The openings between the auricles and ventricles are guarded by little swing-doors, called **valves.** These valves may be roughly compared to folding-doors or gates, which, by opening only one way, allow the blood to flow in that direction, and

**Work done by the Heart.** — The heart is a wonderfully busy machine, pumping away without getting tired, night and day for eighty years or more, perhaps, at the rate of seventy-two strokes every minute, over forty-three hundred times every hour, and nearly thirty-eight million beats every year. At each stroke, each ventricle pumps about six ounces, or nearly fifty teaspoonfuls, of blood. About eighteen pounds of blood are moved every minute, or twelve tons every day.

It is calculated that the total amount of daily work done by the heart in a full-sized man is equivalent to the lifting of a ton weight to the height of two hundred feet. This is estimated to be about one-fifth of the whole amount of energy which a man puts forth in the form of heat and motion. While the cavities are filling with blood, and its muscles are relaxed, the heart has a brief rest: otherwise, it could not keep up its patient and tireless pumping of over four thousand tons of blood every year, from birth to death.

prevent its flowing in any other.    They are operated by slender but powerful muscles within the ventricles.

The valve on the right side of the heart is called the three-pointed or *tricuspid* valve.

That on the left is said to look like a bishop's mitre : hence it is called the *mitral* valve.

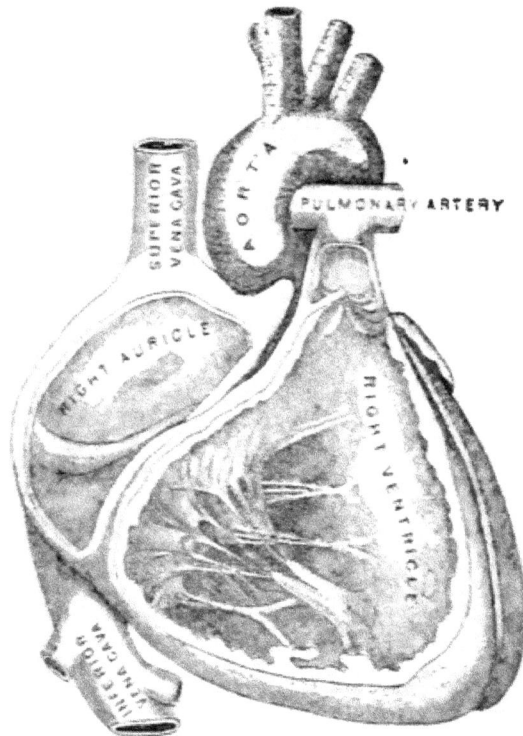

FIG. 43. — Right Section of Heart.

Between the ventricles and the arteries are the *semilunar* valves, so called from their shapes.

The valves fall back to let the blood flow from the auricles into the ventricles, but float up with the blood so as to prevent the return of the blood into the auricles. They are prevented from floating over into the auricles by delicate cords which tie them to the ventricles.

109. **Blood-Vessels connected with the Heart.** — The **aorta,** which is the largest artery in the body, springs from the left ventricle. It carries the bright, pure blood out from the heart.

Four **pulmonary veins** open into the left auricle. Two of these veins come from the right lung, and two from the left lung. They bring back to the heart the blood which has been purified in the lungs. Two of the largest veins in the body, called the superior and inferior vena cava, open into the right auricle. Both of these great veins pour into the right auricle, the dark, impure blood, which has been collected in various parts of the body by the smaller veins.

The **pulmonary artery** springs from the right ventricle. Soon after leaving the heart, it splits into two pipes; one goes to the right lung, the other to the left lung. This artery carries from the heart to the lungs the dark, impure blood, which has been brought to it by the great veins. Its entrance is guarded by the semilunar valves.

110. **The Arteries.** — The **arteries** are the pipes which carry blood from the heart to all parts of the body.

The arteries may be regarded as branches of the aorta, or the main artery, which starts from the left ventricle of the heart. After leaving the heart, the aorta rises towards the neck, but soon turns downwards forming a curve, called the "arch of the aorta." This great pipe, passing between the lungs to the back, then runs down along the spine, through the diaphragm. In tne lower part of the abdomen, it divides into two main branches, one of which goes to each lower limb.[1]

Two large arteries spring from the arch of the aorta, and run up on each side of the neck to the head.[2]  Again there are two large arteries, which branch off from the aorta, pass beneath the collar-bone, and supply branches to the arms.[3]

While the aorta is passing down the spine, it gives off branches to the important organs of the abdomen.

III. **The Veins.** — The veins are the return-pipes that bring the blood back to the heart.

Unlike the arteries, which gradually grow smaller and smaller, the veins, starting from the capillaries, grow continually larger and larger.  The veins of the legs travel upwards, becoming gradually larger by the addition of other branches in the abdomen, until at last all the united branches are joined in one great vein,[4] which empties into the right auricle of the heart.  The venous blood from the head and arms flows back, and empties into the right auricle by another large vein.[5]

Veins generally lie near the surface of the body, just beneath the skin.  We may see them in almost any part of the body.

The veins are abundantly supplied with little pouch-like folds, or pockets, which act as valves, and allow the blood to flow only towards the heart.  If we press the fingers along one of the veins in the arm, towards the hand, we shall see a number of little knots, or

---

[1] Femoral.          [2] Carotid.          [3] Subclavian.
[4] Inferior vena cava.      [5] Superior vena cava.

swellings, here and there along the vein. The blood thus forced back fills the little pockets in the vein. Take away the finger, and the knots will at once disappear, because the blood is left free to flow towards the heart.

**112. The Capillaries.** — Between the end of the smallest arteries and the beginning of the tiniest veins, is a very close network, like the finest lace, with the minutest little tubes for threads. These little tubes are the **capillaries,** or hair-like vessels.

In reality, they are as much smaller than hairs, as hairs are smaller than cables. So closely set are these tiny vessels, that we cannot prick any part of the skin, even with the smallest needle, without wounding one or more of them, and drawing blood. Some of them are so small that three thousand of them placed side by side would not, in their united width, measure more than one inch. The blood corpuscles can only pass through them in single file.

**The Veins compared to a Sewer System.** — As we compared the arteries to the water-supply pipes of a great city, we may now compare the veins to another underground network of pipes, commonly called the "sewer system." The pure, wholesome water is brought by the street pipes to each house. It is used, becomes filthy with dirt, waste, and all kinds of impurities ; then the drains from each house carry it off before it can do any harm. The sewer-pipes from the houses unite with those of the streets ; those of the streets unite at last into one great main sewer, through which flows this unwholesome river of impurities.

The veins of the body resemble this sewer system. The pure, wholesome blood is brought by the arteries, is used by the tissues, and becomes foul with impurities. The veins are the drain-pipes of the body which carry away this impure blood from the tissues, and spread out their contents into Nature's great purifying reservoir, the lungs.

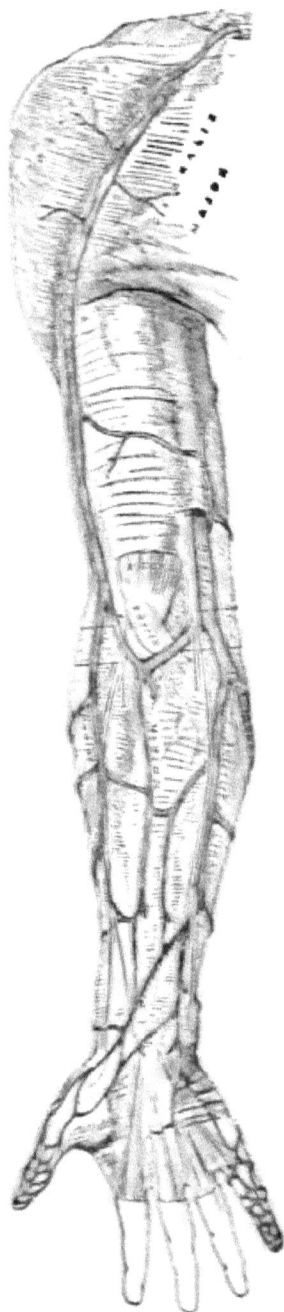

The nutriment brought by the red corpuscles soaks through the delicate, thin walls of the capillaries, and the tissues are bathed with the life-giving fluid of the blood. It is in this way that the tissues of the body are nourished, strengthened, and renewed. At the same time, the worn-out tissues are seized and burned up by the oxygen brought to them by the blood in the capillaries. The products of this burning are then caught up by the blood, and sent along to the veins, and so on to the heart.

**113. Circulation of the Blood.** — Let us begin at the right auricle. We find two large veins busily filling the auricle with dark, impure blood, collected from all parts of the body. As soon as the auricle is full it begins to contract. The blood cannot get back into the great veins because it is flushed forward by the great volume of blood behind it. The door opening

Fig. 30. — The Superficial Veins of the Arm.

into the right ventricle lies open, through which the blood flows until it is full. The ventricle now begins to contract; the tricuspid valve at once closes, and thus prevents the reflow of blood. The blood is driven into the pulmonary artery through the semilunar valve.

This artery carries the blood to the lungs. The dark, impure blood is driven along narrower and narrower vessels until it reaches the capillaries of the lungs. Here it is, as it were, spread out to be purified. Exposed to the oxygen of the air, the blood gives up its impurities and its purple look.

It takes up a great deal of the oxygen of the air in exchange; and in a purified state, and of a bright scarlet color,

FIG. 51. — Diagram of the Arterial System.

it comes back to the heart by the pulmonary veins, which pour it into the left auricle.

From the left auricle the blood is forced through the mitral valve into the left ventricle. As soon as the left ventricle is full, it begins to contract. The mitral valve

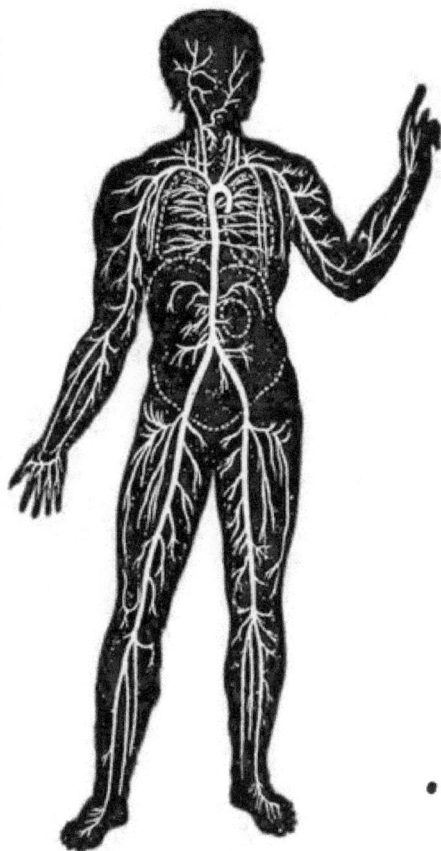

at once closes, and blocks up the passage into the left auricle; and the blood has no other way open but through the semilunar valve [1] into the aorta.

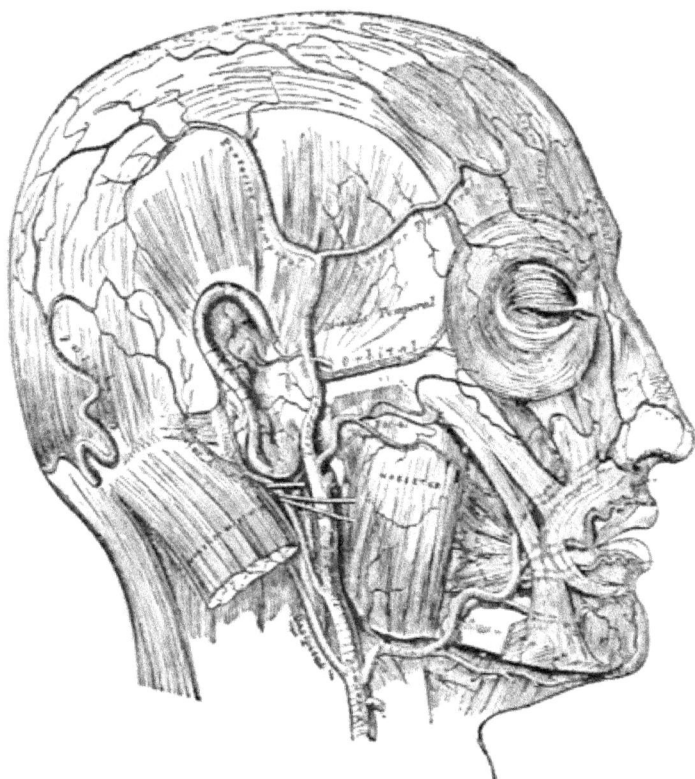

FIG 52. — Arteries of the Neck and Head.

The aorta and its branches, as we already know, distribute the blood through every tissue of the body. From the tissues it is again returned by the veins to

[1] The entrance to the aorta is guarded by a semilunar valve similar to that at the entrance of the pulmonary arteries.

the right auricle of the heart, and thus the round of circulation is continually kept up.[1]

**114. Sounds of the Heart.** — As the heart is doing its work, it gives out certain sounds. If we put our ear against the chest of another person, over the region of the heart, we can easily hear two " pit-pat " sounds, one longer than the other. The first sound takes place when the walls of the heart contract, and force out the blood. The sharp, second sound is heard when the doors, or semilunar valves, shut.

When the ventricles contract, the apex of the heart strikes against the wall of the chest, on the left side, between the fifth and sixth ribs. This is known as the " beating " of the heart, which makes itself felt when unusually strong, as in running and jumping, or during any other severe exercise. Sudden joy or fear will also make the heart beat quickly.

Thus, while we cannot make the heart go slower or faster by any effort of the will, it is greatly influenced by the feelings.[2] It is quickened by mental excitement;

---

[1] A drop of blood goes the grand round of the body in about twenty-two seconds ; that is, while we are counting twenty-two, the blood flows through the right side of the heart, passing out into the smallest capillaries of the lungs, is purified, absorbs oxygen, is picked up again by the veins, and returned to the left side of the heart, to be driven out through the arteries into the capillaries, picked up again by the veins, and returned to the right side of the heart, — thus making the grand round. The entire blood in the body is thus able to make this complete circuit in about two minutes.

[2] People who work hard and are poorly fed, often suffer from an irregular beating of the heart after some exertion, as walking up a steep hill, or going upstairs, or sweeping. The feeling is as if the heart were going to " jump out," and is commonly known as " palpitation." It is purely a disturbance of the functions of the heart, and has in no wise to do with heart disease, from which people die so suddenly. Palpitation of the heart is occasionally caused by long-continued drinking of strong tea and coffee. Tobacco will often bring on the same trouble. Thus we have the " tea-drinker's heart " and the " tobacco heart."

while a sudden shock to the mind may cause such a failure of the heart's action as to cause fainting, and in very rare cases death itself.

**115. The Pulse.**—If the finger be put over an artery which lies near the surface of the body, like the radial artery, for instance, near the wrist, or the temporal artery, just over the temples, a regular throbbing, called the **pulse,** will be felt.

This is the reason for it: when the left ventricle contracts, the blood is forced into the arteries, already full of blood, faster than it can pass through them. The elastic arteries are stretched by the extra blood; and their rhythmical throbbing, keeping time with the heart beat, is felt by the finger.

In a healthy grown person, the pulse beats about seventy-two times a minute. In children the pulse is quicker than in adults, and slower in old age than in middle life. As the pulse varies very much in its rate and character in disease, it is to the trained touch of the physician a good index of the bodily condition of his patient. The pulse is commonly felt at the wrist simply because that is the most convenient and accessible; but it may also be felt at any place in the body when an artery runs near enough to the surface.

**116. Effect of Alcohol on the Circulation.**—The flushed face of the drunkard, or even that of the moderate drinker, is an every-day sight. It may seem to him and to others a sign of health, but it is really one of the many symptoms of alcoholic poisoning. The alcohol has deadened the nerves which regulate the flow of blood. The walls of the blood-vessels stretch, thus allowing more blood to flow through them. The arteries are relaxed, the capillaries are engorged with blood, and the undue amount of blood shows through the skin, making it look red.

This action, — medical men call it "congestion," — we must remember, is not confined to the vessels near the surface of the body, but really extends to every organ and to every tissue.

**117. How Alcohol gets into the Blood.** — Alcohol passes into the blood by two distinct routes. When we take alcoholic liquor into the stomach, some of it at once soaks through the coats of the tiny blood-vessels with which the lining of the stomach is covered, and is carried into the blood-current by the portal circulation.

Alcohol is also taken up by the lacteals, and is emptied into the blood through the thoracic duct. Now, although alcohol goes to the heart, to be sent to every part of the body in a roundabout way through the liver, it takes only a moment or two for it to get into the main blood-stream. A glass of strong drink soon "goes to the head," as many people know ; showing that its effects are rapidly produced in the remotest tissues of the brain. The rapidity of this absorption depends upon the kind of liquor and the condition of the stomach.

Alcohol thus taken into the blood is flushed everywhere, — into each fibre, membrane, and tissue. It saturates all the vital organs, of the brain, heart, liver, lungs, kidneys, skin, and the secreting-apparatus.[1]

---

1 " The habitual use of alcohol produces a deleterious influence upon the whole economy. The digestive powers are weakened, the appetite is impaired, and the muscular system is enfeebled. The blood is impoverished, and nutrition is imperfect and disordered, as shown by the flabbiness of the skin and muscles, emaciation, or an abnormal accumulation of fat." — DR. AUSTIN FLINT.

**118. Effect of Alcohol upon the Blood.** — If a sufficient quantity is present it shrinks and hardens the tiny corpuscles, rendering them no longer able to carry the oxygen the blood needs to keep it pure. Waste matter and fat, that should be burned up by the oxygen to keep the body warm, circulate in the blood, making it impure. This fat may be deposited in the muscles, heart, and other places where it is not needed, and may slowly crowd upon and take the place of the tissues of which those organs are composed. The fat cannot perform the work of the tissue it displaces, and disease is the result. The blood of the beer-drinker becomes unduly thin, and destitute of the material for forming a clot in case of wounds. Slight cuts may result in dangerous or fatal bleeding.[1]

**119. Effect of Alcohol upon the Heart.** — The paralyzing effect of alcohol on the nerves which regulate the flow of blood in the arteries causes them to lose their grip on the blood-vessels, and hence the heart has less resistance to meet. Consequently it is forced to beat faster, and work harder, to fill the dilated vessels with blood.[2] The alcohol paralyzes other nerves whose office is to hold the heart in check and to keep it from beating too fast. Its effect is like that of taking the weight off the pendulum of a clock : it beats wildly and rapidly, without anything to control it. This increased frequency of the heart's stroke results in increased labor, more wear and tear, and less rest for this vital organ.

[1] See Note 15, page 371.

[2] The late Dr. Parkes, an eminent English authority on matters of health, calculated that this extra work of the heart, when alcoholic liquors were consumed, amounted to the lifting of twenty-four tons of extra blood one foot high each day, — about twenty per cent of additional work.

It has been estimated that the amount of extra work which alcohol puts upon the heart is equal to the work of lifting from five to thirty tons, according to the amount taken, one foot high in the course of twenty-four hours. This effect of alcohol is often called its stimulant action. It is more exact to call it the beginning of the poisonous action of alcohol.

Here is the whole story : —

Alochol causes an increase of action by narcotizing the nerves which regulate the heart.[1]

**120. Effect of Tobacco on the Heart.** — Tobacco causes irregularity and palpitation of the heart, frequently accompanied with attacks of severe pain. Physicians report many cases of heart disease among young boys, brought on by the use of tobacco.[2] A celebrated physician says it invariably produces a weak, tremulous pulse. A German physician says that in his country, while even immoderate smoking may appear to agree with persons for many years, suddenly and without any other assignable cause, trouble with the heart begins. Leading physicians in our own country agree that tobacco is the true cause of a great proportion of the fatal cases of heart disease.[3]

1 See Note 16, page 373.

2 " Tobacco, and especially cigarettes, being a *depressant upon the heart*, should be positively forbidden." — DR. J. M. KEATING, on " Physical Development," in *Cyclopædia of the Diseases of Children.*

3 See Note 17, page 374.

# CHAPTER IX

## BREATHING

**121. Why we Breathe.** — Night and day, without one moment's rest, from the first to the last instant of our lives, we are breathing. About eighteen times every minute, more than twenty-four thousand times every day, **and** many millions of times in a lifetime of seventy years, we draw in and send out again our breath.

Most of the time we do not think anything about **it**. We eat, talk, work, and sleep; **and** all this time our breathing goes quietly on. Breathing is the plainest sign by which **we judge** a person to be alive who lies otherwise motionless and insensible. Though we can hold our breath for a short time, **yet, after a** few seconds, we begin to feel uncomfortable.

In speaking of food and the blood, we have learned, that, without food and fresh air, the burning, or oxidation, which is slowly going on in our bodies all the time, would soon flag, and we should, after awhile, die for want of them. Hence we must have oxygen to keep the burning. We get the oxygen from the air we breathe.

Again: we have learned that the dark, impure, and venous blood was sent to the lungs from the right ventricle, and was returned to the left auricle as pure **arterial** blood of a bright red color. The blood **has got rid of its** waste matters, and becomes pure. **In brief,** the **blood has** been oxidized in the act of breathing.

Hence in breathing we have two objects in view, —

First, to bring a fresh supply of oxygen to the blood, by means of the lungs.

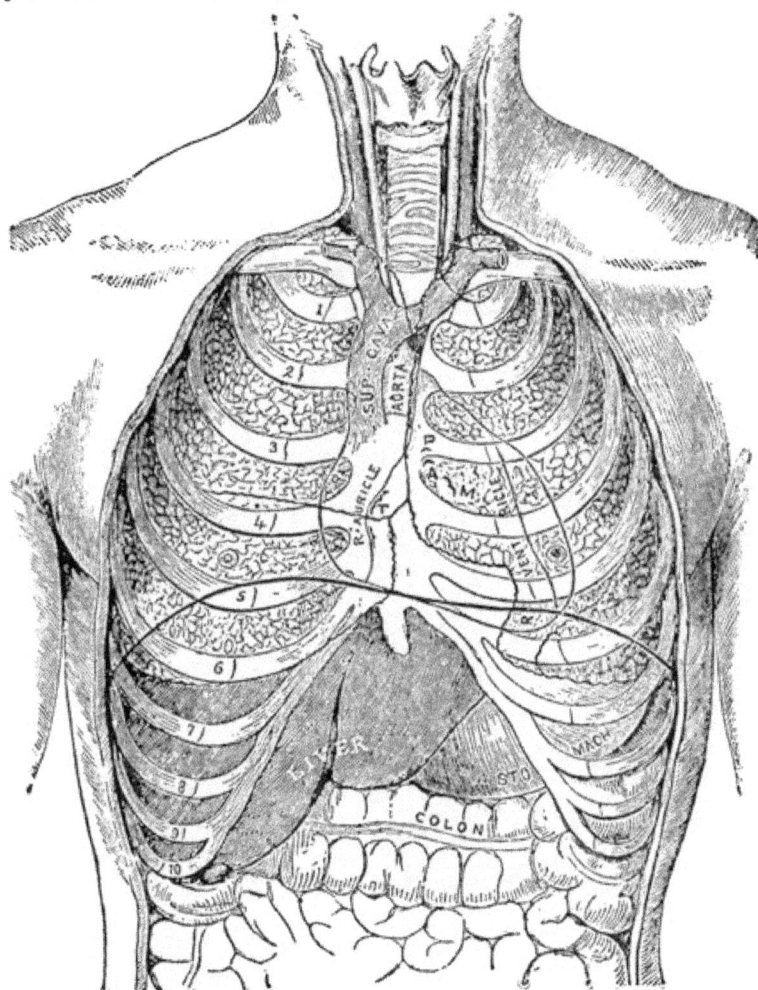

FIG. 53. — Showing the relative position of the Heart, Lungs, Liver, etc., with reference to each other and the chest-walls.

Second, to get rid of carbonic acid and waste matters taken up from the tissues, and brought to the lungs by the blood.

**122. The Air-Passages.** — The air is drawn into the lungs through the mouth, nostrils, and windpipe.

The **nostrils** are really the passage-ways for the air, and warm the air somewhat before it goes into the lungs. The air passes from the nostrils into the windpipe, or trachea.

The **windpipe** is a hollow tube about four inches long, and is protected on the front and sides by stout rings of gristle. But for them the windpipe would close with the slightest pressure, and cut off the breath. The top of the windpipe is protected by a trap-door known as the *epiglottis*. When food is swallowed, this little door shuts tight, and keeps it out of the air-passages: otherwise the food would go the wrong way, and cut short our breath. When we breathe, the lid lifts up.

The upper part of the windpipe is a kind of box called the **larynx,** or organ of voice. In this box — which in some persons is very prominent, and the front of which is commonly called "Adam's apple" — are the vocal cords. These cords are not strings, but rather elastic strips, with free edges which can be made tight or slack. As the air goes to and from the lungs, through the narrow slit between them, called the *glottis*, it sets the cords to vibrating, and thus makes the sound of the voice.

The windpipe, after entering the chest, divides into two branches called **bronchi,** or windpipe tubes, sending one to each lung. These, again, divide into smaller tubes called **bronchial tubes.** Each bronchial tube divides again into smaller branches, these again into smaller, and so on to the tiniest twigs, a hundred times smaller than the hairs of our head.

If we only remember that all these tubes, great and small, are hollow, we may compare the whole system to a short bush or tree growing upside down in the chest, of which the windpipe is the trunk, and the bronchial tubes the branches of various sizes.

123. **The Lungs.** — The lungs are two large, pinkish, spongy organs, which surround the heart, and fill up all the rest of the chest-cavity.

The right lung is the larger of the two, and has three parts, or *lobes* The left lung has only two lobes. The inside structure of the lungs is really a mass of air-passages, arteries, veins, and capillaries.

Imagine a thick, short tree crowded with leaves : imagine the trunk and

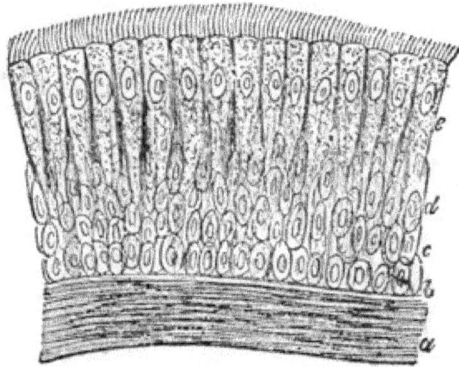

FIG. 54. — Ciliated Epithelium from the Human Trachea (magnified 350 times): *a*, innermost layers of the elastic longitudinal fibres; *b*, innermost layers of the mucous membrane ; *c*, deepest round cells ; *d*, middle elongated ; *e*, superficial, bearing cilia.

**How Nature protects the Air-Passages.** — The inside walls of the windpipe and air-tubes are lined with a mucous membrane, which secretes a thick, sticky fluid called mucus, to keep the passages moist. This membrane is covered with thread-like processes called cilia. They are not unlike fine hairs, and resemble somewhat the pile on velvet. They wave to and fro, like a field of grain in the breeze.

The motion is always upwards and outwards towards the mouth. They make one continuous brush, which is ever busy sweeping out the air-tubes. They sweep up the particles of dust and mucus from the tubes, and brush them out of the pipes with a sudden blast of air which we call coughing. These tiny cilia are simply the dusters which Nature uses to keep the air-passages neat and clean.

branches, small and great down to the smallest twigs, are
hollow.   Suppose the leaves themselves were tiny blad-
ders, blown up and tied on to the smallest hollow twigs,
and made up of some delicate but very elastic substance.

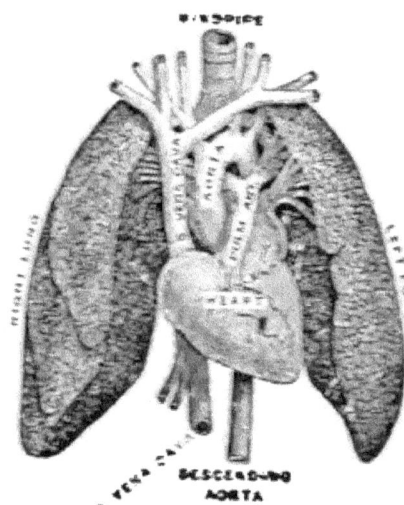

Fig. 33 — The Lungs and Heart, viewed in front.

Around such a frame-
work of hollow branches,
called bronchial tubes,
and hollow, elastic blad-
ders, called air-cells, is
wrapped a finely woven
network of arteries or
veins, and capillaries.
Imagine a child's ball
covered with a fine net-
work of red and blue
yarn.

We can form some idea
of the number of the air-
cells when we remember that the end of the smallest
bronchial tube, often called a lung-sac,[1] is said to hold
about seventeen hundred air-cells.  The lungs, then,
consist of millions of these tiny air-cells, packed closely
together, each air-cell communicating with the windpipe, and so with the atmosphere in which we breathe.

The chest is lined, and the lungs covered, with a
smooth, delicate lining, called the *pleura*.  These two
surfaces rub against each other when we breathe.

[1] Each lung-sac bears some resemblance to a bunch of grapes, and is so small
that forty of them would barely reach an inch.  Just as the bunch of grapes, however, contains a large number of single grapes, so every lung-sac consists of a
large number of separate air-cells.

This lining secretes a fluid which keeps the parts always moist, and prevents their rubbing one on another.

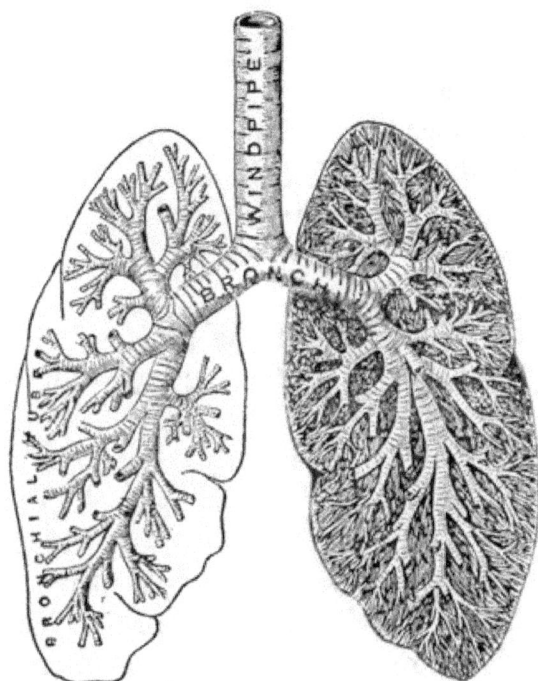

FIG. 56. — Section showing the ramifications of the Bronchi in the Lungs.

**124. How We Breathe.** — If we put both hands on the sides of our own chest, and breathe in deeply, we feel that the act has carried the hands farther apart. Again, if we put one hand across the middle of our chest, we feel that it is carried forward every time we breathe in, and is returned to its place as we breathe out.

Further, if we, pass a tape-measure round the chest, and draw it tight when we breathe out, we find that the tape must be let out two or three inches when we begin to breathe in. If we breathe in and out with a great deal of force, the changes are more marked.

Hence there are two movements in breathing, — one in which the cavity of the chest is made larger in all its dimensions : this is when we breathe in air, and is called **inspiration.**

The other movement is the one in which the chest-cavity is made smaller in like manner : this is when we breathe out air, and is called **expiration.** The means by which the air is breathed in and out of the lungs makes up the act of breathing.

**125. The Act of Breathing.** — The cavity of the chest is a closed air-tight chamber, whose only opening is the windpipe. The pressure of the air in the air-passages keeps the lungs stretched out so as to fill this cavity. Imagine now such a chamber as this to have a kind of false floor, capable of moving up and down. When the floor was crowded down, the cavity would be enlarged ; and the pressure of the air would then cause the elastic lungs to expand to a greater extent to fill up the extra space.

When the floor was raised again, the cavity would be diminished ; and the stretched lungs, being diminished also, would give up the extra quantity of air which they took in.

Suppose these two movements were performed at regular intervals. Then, every time the floor was pressed down, there would be a rush of air down the air-passages to the lungs ; and, every time it was raised, that same quantity of air would be driven out of the lungs again.

This is exactly what happens during the process of breathing. The bottom, or floor, of the chest, is formed, as we know, by a large, flat muscle called the diaphragm.

When this muscle contracts, it is pressed down, and the cavity of the chest is enlarged. When it relaxes, and consequently rises again, this chamber is diminished. These two up-and-down movements of the diaphragm are the chief movements in ordinary, quiet breathing.

The cavity of the chest, however, is enlarged in another way. The walls of the chest are formed by the ribs, which encircle it so as to join the breastbone in front. The spaces between the ribs are occupied by a set of strong muscles called the intercostal muscles.

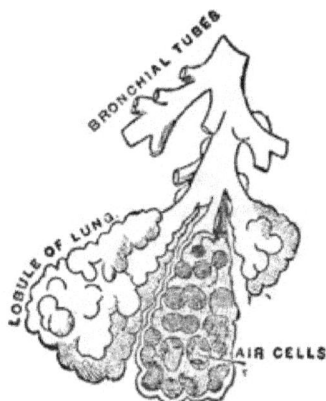

Fig. 57. — The Air-cells and Bronchial Tubes of the Lungs.

One set of these intercostal muscles contracts, and pulls up the ribs, which are fastened to the backbone behind by a joint. When the ribs are raised, they push out the breast-bone in front ; and thus the cavity of the chest is enlarged all round. This enlargement by means of the side walls takes place at the same time as the diaphragm descends, so that the chest is enlarged on all sides. An extra quantity of air then rushes into the lungs, and we get an **inspiration.**

Immediately following the inspiration, the diaphragm relaxes, and, of course, rises ; and, at the same time,

another set of intercostal muscles begins to pull the
ribs and breast-bone down. These combined movements diminish the cavity of the chest, and consequently the same quantity of air is driven out. This
makes an **expiration**.

126. **Changes in the Air from Breathing.** — Air is a
mixture of two gases, — *oxygen* and *nitrogen*, — in the
proportion of one part of the former to four of the latter. Oxygen is the active gas, the feeding and warming
gas, the life-giving principle of nature. It has been
well named " the great supporter of animal life."
Nitrogen is mixed with it, lest the oxygen should be
too strong for us, and burn us away too fast. In short,
Nature dilutes the oxygen of the air for us with nitrogen.

If we examine the air just as it enters the lungs, and
again after it has passed through them, we shall find
that, while the bulk is almost exactly the same, the
quality has been changed. It has left behind about
one-twentieth of its oxygen, and taken in exchange for
it nearly the same quantity of carbonic acid. About
twenty cubic inches of air pass in and out of the lungs
with every breath, and about three hundred cubic feet
every twenty-four hours ; amounting to the contents of
sixty barrels.

Let us try to understand how this interchange takes
place between the air and the blood in the lungs.

Experiments carried on outside of the body prove
that gases can pass through delicate membranes. If
a bladder is filled with oxygen, and then hung up in a
bottle filled with carbonic acid, the two gases will mix
with each other. The oxygen will pass out through the

walls of the bladder, and the carbonic acid will pass in. This is in accordance with a well-known law of physical science.

This is practically what happens in our lungs every moment of our lives. The blood and the air are separated only by the thin and delicate walls of the air-cells, and by the walls of the capillary blood-vessels of the lungs. The matters contained in the blood pass outwards into the lung, whilst the matters contained in the lung pass inwards to the blood. This last-mentioned act constitutes the essential feature in the function of respiration. The blood, thus renewed, travels to the left side of the heart, is pumped out through the aorta, and distributed to every tissue of the body.

We may, in brief, look upon the lungs as a kind of market-place or exchange, where two merchants — the blood and the air — meet to exchange their wares. Indeed, it is a very busy market-place.

This, then, is the whole story, shortly told, of our constant need of air: The tissues of the body, of whatever kind, everywhere over the body, breathe blood, making pure arterial blood venous and impure, in every part of our bodies, except in the lungs, where the blood itself breathes air, and changes from impure and venous to pure and arterial.

The air, as it leaves the lungs, is saturated with watery vapor. This is seen when we breathe on the bright steel blade of a pocket-knife, a mirror, or any cold, polished surface. As we all know, the surface becomes covered with a thin film, or minute drops of water. In cold weather this moisture becomes visible with each expiration.

Air as it leaves the lungs is much hotter than the surrounding air. It is generally about 98° F. For this reason, on a cold day we blow our fingers to warm them. The air breathed out of the lungs also contains a small amount of decaying animal matter. Every one knows the unpleasant odor of the air in rooms in which many persons have been closely shut up.

**127. Carbonic Acid and Its Poisonous Effects.** — Carbonic acid, in its pure state, acts as a deadly poison. An excess of it in the air produces poisonous effects. The air we breathe out contains four parts in a hundred of this gas. Increase this to ten parts, and it will prove a deadly poison to warm-blooded animals. In smaller quantities this gas produces labored breathing, dizziness, headache, and a general stagnation of the bodily life.[1]

In the mines this gas becomes the dreaded "choke-damp." It is dangerous oftentimes to cross over a vat in which beer is actively fermenting; for the air over the vat, loaded with carbonic acid, is utterly unfit to breathe.

In the open air we rarely suffer any ill effects from carbonic acid, for the simple reason that Nature is always mingling the gas with the oxygen by means of the winds and the rains. A fresh supply of pure,

---

[1] This same deadly gas sometimes flows naturally from the earth. Some of us have heard of the " Valley of Poison " in the island of Java, of which fabulous stories are told by travellers. This gas is very abundant in this valley, and from its weight sinks to the ground, where may be seen, it is said, the skeletons of birds and animals, which have been suffocated in their attempt to cross this death-trap. There is a lake in Italy called Avernus, meaning "without birds." As the story runs, birds, in flying over the lake, are poisoned by the carbonic acid, and fall dead into the water.

life-giving air is thus furnished in the greatest abundance.

Like a prudent manager, Nature utilizes the carbonic acid as the life-food of the vegetable creation. The trees, the plants, and the grasses make, as it were, oxygen out of the carbonic acid; while we make carbonic acid out of the oxygen taken into the lungs.

**128. Impurities in the Air.** — While carbonic acid and the waste animal matter given off with it from the lungs, are the most common impurities, there are many other things which make the air unwholesome. The poisoned air due to cesspools, drains, and sewers, is often a frequent source of disease.

Sewer-gas is a fruitful source of certain diseases, as typhoid-fever and diphtheria. The foul air from chemical works, bone and soap factories, and many other manufacturing places, is more or less hurtful to health. Even the dust in our rooms may carry, it is thought, the germs of disease.

Certain occupations may shorten life by exposure to air loaded with impurities. Thus, there is the "miner's consumption," or "black lung," due to the dust breathed into the lungs, acting like so many little splinters in the delicate air-cells. Those who work on steel, emery, pottery, etc., also suffer from the irritating dust floating in the air. Other impurities are highly injurious to the lungs, as the dust in match-factories, white-lead works, copper and brass founderies, and arsenic in wall-papers.

Unwholesome as the air may be in the workrooms of many trades, the real danger, after all, and that which

should be of more concern to the public generally,
arises from the slow and unsuspected effects of breathing air which has already been breathed. The peril is
in our own living-rooms, our bedrooms, our schoolrooms, halls, vestries, theatres, and churches.

129. **How to Ventilate.** — The best way to rid the
air of its impurities is by some suitable system of **ventilation.** To ventilate a place is to cause pure air to
flow through it. In other words, it is some practical
plan to keep the air pure and wholesome. Do our best,
and we cannot keep the air of any inhabited room as
pure as the atmosphere outside.

The object aimed at in ventilation, is to give an outlet to impure air, and an inlet to that which is pure,
fresh, and moist. Remember that it is not at all necessary that air should be cold to be pure. The required
amount of fresh air should be moved evenly through
the room or building with a gentle current, and without a draught.

An open fireplace is a healthful, safe, but not economical means by which to heat and ventilate a room.
Stoves in a room soon dry the air, unless fresh quantities from outside are constantly supplied. When rooms
are warmed by heated air from furnaces, the warm air
should enter through registers in or near the floor, on
one side of the room; and impure air should escape
through outlets in or near the ceiling, on the other side.

Children should be trained from infancy to sleep
with the windows partly open for the greater part of
the year. Adult people in vigorous health should
gradually learn to do the same. Even in the coldest

weather, some simple apparatus to let in fresh air is as good as any. Raise the window a few inches, and put in a piece of board under the lower sash. ·Pure air will enter where the two sashes overlap.

A common window-screen, cut to the proper size, and covered with flannel instead of wire, will let in plenty of air without draught, and is suitable for cold weather. Again, fit an elbow of common stove-pipe into a board of the right size, and put it under the raised sash, with a damper to regulate the current of air. These and other contrivances are easy to make, and cost but little. They answer the purpose even better than those that are more costly and complicated.

130. **Ventilation of Schoolrooms.** — Special pains must be taken to ventilate schoolrooms. Pupils are sure to be listless, uneasy, dull, and sleepy when the air is not wholesome. Children may be comfortable in a well-aired room at 66° F., but it is very easy to let the temperature run up to 85° before it is noticed. Whatever the apparatus for ventilation may be, the doors and windows should be opened before and after each session and at recess.

The air of the room should be changed as often as once every hour. The pupils meanwhile should engage in active gymnastic exercises to prevent taking cold. When this is done in cold weather, the heat should be turned on so as to warm the cold air coming in as quickly as possible.

Weakly children, those liable to croup, those easy to catch cold and other ailments, must be carefully looked after. Never allow draughts of cold air to fall directly on the heads of children. Guard the air of the school-room from the foul air arising from closets, outbuildings, sinks, cesspools, and all other possible sources of ill-health.

**131. Why the Body is Warm.** — Every one knows that the surface of the body feels warm. Hold the fingers in the mouth, and we find it warm. Put a thermometer, made for the purpose, in the armpit, or in the mouth for five minutes, and it will register about 98° F., even in the coldest day of midwinter. This is the natural heat of a healthy person; and it rarely varies more than a degree or two, except in disease.[1]

This heat is produced in just as simple a manner as that which results from a common fire or a lighted candle : it is the natural result of the process of combustion. We are warm ourselves, because we are burning away bit by bit, just as a candle does; that is to say, by the union of carbon, or charcoal, with oxygen. There is only this difference : we burn wet materials (the moist tissues), and do not make a flame or give a light. We take our coal or tallow in the shape of starch, sugar, and fat, and get the oxygen from the air we breathe.

A steam-engine at work is warm because all the energy set free from the fuel burned is not turned into mechanical work, but some of it appears as heat : so it

[1] The rate of combustion may be much increased or lessened in various diseases. In pneumonia, typhoid-fever, and blood-poisoning, the physician may note a temperature of 105° F., or even more : hence the fever-patient says repeatedly that he is "burning up" with fever, and eagerly drinks all the cold water which is given him, and calls for more in a few minutes. In other diseases, such as cholera, there is a notable fall in the natural temperature.

is in our bodies. Our muscles, our organs, in fact, every tiny cell, is busily at work ; and their substance is slowly burned at a low temperature. Every time we move, feel, think, or exercise any function, this oxidation, or burning, goes on.

Some of the energy thus set free by this slow combustion shows itself as heat, which helps keep the body warm and at its natural temperature. Thus, animal heat is produced, and life maintained. Our bodies are working at a temperature higher than the surrounding air, except in the hottest weather : hence there must be a loss of heat nearly all the time. Therefore we must keep making heat all the time to compensate for the continual loss.

Besides this loss by radiation, as it is called, considerable moisture is got rid of by the skin in the form of vapor, or sweat. The evaporation of this moisture from the skin acts as a kind of regulator to keep down the excess of heat.

**Proper Temperature for our Living-Rooms.** — The temperature in our living-rooms should be kept at about 68° to 70° F. Most of our rooms are apt to be overheated or unequally heated, especially during cold weather. Any person, child or adult, may become tender and delicate in a short time by getting used to overheated rooms. The temperature of the sick-room depends somewhat upon the age of the sick person, and his disease. A temperature of 40° to 50° F. may be suitable in typhoid-fever, while 80° may be necessary for an aged person suffering from rheumatism.

The greatest care in keeping any sick-room supplied with pure, fresh air, and at the proper temperature, is necessary. Old people generally need a higher temperature than the young and vigorous. The warm air in our living-rooms, to be wholesome, should be kept moist. A shallow pan of water put on the stove, or near a register, answers every purpose.

**132. Effects of Alcohol upon the Lungs.** — The use of drinks containing alcohol tends to bring on inflammation of the lung-tissues, and hence lessen the breathing-capacity. The tissues of the lungs become thickened and hardened by the alcohol, and hence do not allow oxygen to pass through them into the blood, and carbonic-acid gas and other waste matters to pass out as they should. The apparatus called the "spirometer," used by life-insurance companies to test the breathing-capacity of lungs, often detects the dram-drinker by his failure to reach the natural breathing-capacity. The wheezy, broken speech of the drunkard is partly due to this condensation of the lung-tissue.

Again, the repeated dilatation of the lung-capillaries tends to make the habitual user of alcohol more liable to attacks of severe cold, pleurisy, and pneumonia, after making due allowance for the exposure to cold and damp, so common with the intemperate.[1]

**133. Alcohol and the Bodily Heat.** — Soon after taking even a small quantity of alcohol, there is a general feeling of warmth over the surface of the body. The body is not really warmer, but the skin feels warmer. On the contrary, we are really colder, because heat is more rapidly lost by radiation and evaporation from the surface.

The skin is warmer after taking alcoholic liquor, because the nerves that regulate the hair-like blood-vessels on the surface, being partly paralyzed or deadened,

---

[1] According to some good medical authorities, the use of alcohol works such a change in the lung-tissues as not unfrequently to lead to a form of consumption called "alcoholic phthisis."

stretch and let more blood run through them. Hence more blood is sent from the central parts of the body to the surface. There is no real increase of heat : the surface is warmed for the time at the expense of the inner and deeper portions of the body. This surface warmth is now rapidly lost by radiation, and the general heat of the body is lowered below its natural temperature. The bodily temperature is regulated by the surface circulation ; and when this control is lost, as it is by alcohol, the body is cooled by the undue amount of blood carried to the surface.

Experience has proved, time and time again, that alcohol lessens our power to endure extremes of heat or cold for any great length of time. Arctic explorers strictly forbid the use of alcoholic liquor among their men. It has been proved that the exposure to severe cold can be endured far better without alcohol. So well is this bad effect of alcohol known by the people of the coldest regions of Canada, that they will seldom take even a single glass of spirits when exposed to severe cold.

Army life is perhaps the best possible test. It is the almost universal experience of British army officers who have led their men through the recent campaigns in the hottest parts of the Soudan, and who have given special study to the question, that alcohol, so far from being an aid to endure severe exertion and to resist great extremes of heat, acts as a positive injury.

The notion that a dose of some alcoholic liquor taken after exposure or bathing will prevent one from taking cold, is erroneous. The alcohol, by irritating the delicate lung-tissues and lining of the air-passages, and reducing the temperature of the body, makes one more liable to colds, coughs, pneumonia, etc. When one has been chilled, the best thing to do is to get thoroughly warmed, as quickly as possible, either by active exercise or artificial heat.

**Alcohol in Hardship and in Extremes of Heat or Cold.** —" It was quite remarkable to observe how much stronger and more able to do their work the men were when they had nothing but water to drink." — Sir John Ross *in an account of his Arctic explorations.*

" A soldier was given a certain amount of work to do, first when his system was free from the effects of alcohol, and second when under the influence of certain measured doses of brandy. The result is summed up as follows : The brandy seemed to give him a kind of spirit which made him think he could do a great deal of work, but when he came to do it he found he was less capable than he thought. The experience of this man harmonizes with the advice that is given by guides and others who are in the habit of ascending mountains. Spirits, they say, take away the strength from the legs, and should therefore be avoided during a fatiguing expedition."— F. W. Pavy *in "Food and Dietetics."*

The Army of the Potomac, in the spring of 1862, was subjected to great hardships in labor, and exposed to the extremely wet and malarious region of the Chickahominy. There was consequently much sickness and suffering. Under these circumstances the commanding general issued an order allowing every officer and soldier one gill of whiskey per day, half to be served in the morning and half in the evening. The results were so manifestly injurious to the sanitary condition of the army that in just thirty days the order was countermanded by the same general. Concerning this experiment. Dr. Frank Hamilton, one of the most eminent surgeons serving with that army, said, " It is earnestly desired that no such experiment will ever be repeated in the armies of the United States."

# CHAPTER X

## HOW OUR BODIES ARE COVERED

**134. The Skin and How it looks.** — The skin is the outside covering of the body. .

The parts underneath are very tender and sensitive. We all know how painful and tender is any part of the body when the skin has been torn, cut, blistered, burned, or otherwise hurt by accident. Kind nature has given us a strong, elastic, and tight-fitting outside garment. It is easily kept clean, and never wears out. It is soft and thin, yet strong enough to enable us to come in contact with objects without pain or inconvenience.

The skin is richly supplied with nerves and blood-vessels, so closely netted together that it is next to impossible to prick the skin anywhere with the point of a needle without drawing blood, and feeling pain.

**135. The Scarf-Skin.** — The skin is made up of two layers. The outer one has neither blood-vessels nor nerves, and is called the **scarf-skin, cuticle,** or **epidermis.**

The lower layer is called the **true skin,** or **cutis,** which is richly supplied with nerves and blood-vessels. It is so highly sensitive, that, were it not for the scarf-skin, we could not endure life. Most of us are familiar with the delicate pink skin, very sensitive and very painful, which is exposed when the lowest layer of the outer skin is removed by a blister, or rubbed off by some slight accident. The surface feels raw, and oozes a little clear fluid, or perhaps a little blood.

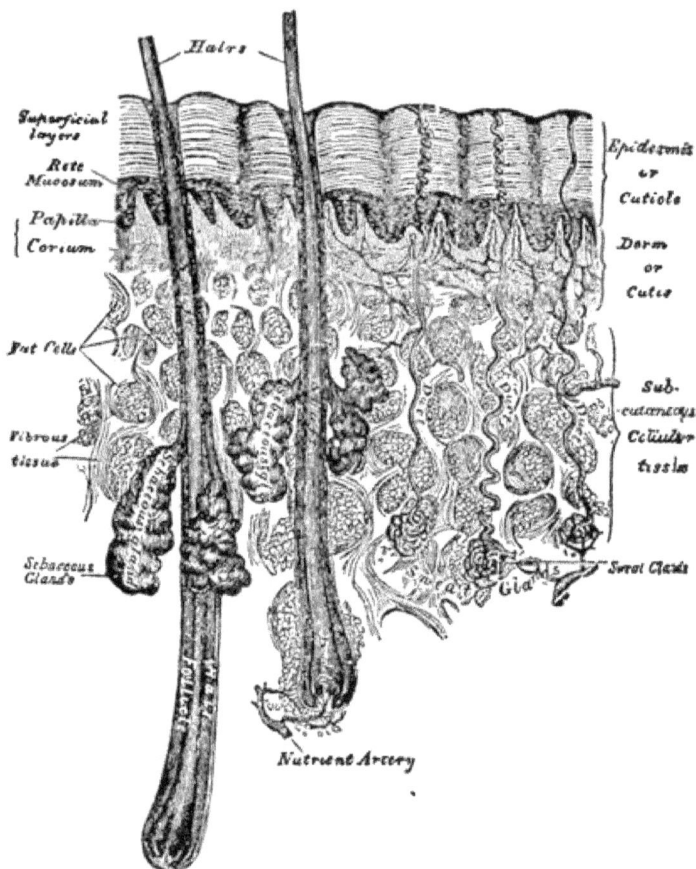

FIG. 58. — A sectional view of the Skin (magnified).

This is the deeper portion of the scarf-skin, which is constantly growing, and developing millions of little round cells to take the place of the flat, horny, and lifeless scales of the outer portion, which are continually dropping off, or being removed by friction.

When these flattened scales are pressed together, they become flatter and flatter ; and thus the hard, horny skin is made, which is seen on the hands of those who use them in hard manual work. The "callus" on the hands of a blacksmith, carpenter, or washerwoman is a familiar sight. Undue pressure or friction from poorly-fitting or tight-fitting boots and shoes causes the hard bunches on and between the toes which are commonly known as "corns."

In the deeper parts of the scarf-skin are tiny specks of coloring-matter, hid in little cells. It is this part of the skin that gives it its color, commonly called the complexion. In the fairer races, these specks are of a pinkish color : in the dark races, the pigment-cells are brown or nearly black, and more closely crowded together. The heat of the sun acts to darken these color-specks, as is seen in the parts of the body exposed to direct sunlight.

We see every day the sharp line drawn between parts of the arm or neck exposed to the sun's rays, and other parts generally covered with clothing. Some, however, tan much more readily than others. When the pigment matter changes in spots, we call them freckles. There may be other defects, such as liver-spots, moth-spots, and other blemishes so often found on the skin. In slight burns, bruises, cuts, blisters, cold-sores, and many eruptions on the skin, the epidermis only is affected. Such injuries and diseases, therefore, heal without a scar.

136. The True Skin. — The true skin, or cutis, is a firm, elastic tissue, resting on meshes of texture some-

thing like damp, raw cotton, which loosely fasten the skin to the parts beneath. It is the true skin which becomes filled with water in dropsy, and which, in the lower animals, is made into leather by the process of tanning. In this layer also are the sweat and oil glands, the hairs, nerves, blood-vessels, and absorbents of the skin.

The outer surface of the true skin rises into little ridges called "papillæ," into which the capillaries and nerves are distributed. These papillæ are very numerous everywhere, but are the thickest where the sense of touch is most acute, as on the tips of the fingers, and on the nose. They are arranged in rows, like hills of corn, and are plainly seen with a magnifying-glass on the palms of the hands.

When the true skin is destroyed, a scar results. White scars, especially on the hands, due to deep cuts, are common enough. Scars from small-pox, deep burns, and other injuries to the true skin, are often seen. The skin is rich in its blood-supply. The nerves are also very abundant. The prick of a pin, or the sting of the smallest insect, causes pain.

**How the Skin may absorb Poison.** — The scarf-skin protects the skin from poisons. Lead, mercury, and other injurious substances will not enter the blood, and affect the bodily health, unless they are actually rubbed through the scarf-skin; but if there is a scratch or sore, so that the true skin is exposed, the poisons may be absorbed into the blood with great rapidity. Workers in lead, looking-glass silverers, and phosphorus-match makers ought, therefore, to take great care to cover the smallest scratch upon their hands. "Lead colic" and "wrist drop" are familiar instances of lead-poisoning. Cheap underclothing, as colored stockings, are often dyed with preparations of lead. Such articles should be thoroughly washed before they are worn. Many hair-dyes contain lead, and often cause lead-poisoning.

**137. The Hair.** — The hair and nails are simply portions of the outer layer of the skin altered in shape and structure. A hair is a slender thread of scarf-skin, which grows from little sacs in the true skin called " hair-follicles." Every hair has two parts, the root and the free end. The root is somewhat pear-shaped, and is sunk in its sac, or follicle, like a post into the ground. In the bottom of this sac is a little hair-papilla (quite distinct from the papillæ of the skin), from which material for the life of the hair comes. As long as this papilla is not destroyed, the hair will grow. Pull out the hair " from the roots," and it will grow again. Destroy this papilla, and the hair never grows again.[1]

One or more little glands open into each hair-sac, and pour out an oily matter for its nourishment. The outside of the hair is quite firm : the inside is softer, and carries the nourishing fluids. Hairs grow from cells pressed together lengthwise, so as to be drawn out into fibres, instead of being flattened into scales. Hence they grow only in length. On the outer surface the cells form a sort of bark, overlapping each other something like the shingles on a roof. The coloring-matter is contained in these cells. When it fails, the hair turns gray or white. The hair-follicle is well furnished with nerves, hence it hurts to have the hair pulled.

---

[1] It is useless, or worse than useless, to try to rid one's self of unsightly hairs or hair-moles on the face. Pull them out with tweezers, cut or shave them off, and they are sure to grow again, coarser and more unsightly than before. All remedies advertised to remove superfluous hairs are worthless or dangerous. The hair-papillæ must be destroyed to stop the growth of hair, and this is no simple matter.

The hairs, or rather the skin close to them, are provided with the tiniest muscles. They run from the
bottom of each hair-follicle in a slanting direction, and
end in the outer part of the true skin. When they
contract they cause the hair to stand more or less erect,
and the skin to " bunch up " a little. Thus, at the sight
of a dog, the hairs on a cat's back become erect and
bristling. Any one who has been frightened suddenly,
or has taken a chilly bath, knows what it is to have
" goose-flesh." [1]

These muscles of the skin also act to squeeze oil out
of the oil-glands. Hairs of some sort are found all over
the surface of the body, except the palms of the hands
and the soles of the feet.[2]    They serve to protect from
heat and cold the parts they cover. On the head, the
hair helps to protect the skull from injuries, and the
brain from extremes of heat and cold. On the body,
they also help to drain off the sweat.

138.   The Nails. — The nails are only portions of the
scarf-skin in a hardened form.

They grow from roots which are lodged in a groove
of the skin, something as a watch-crystal is fitted into
its case. The part which is beneath the skin towards
the hand is called the root, and the rest the body. The
nail rests upon a bed, called the matrix, to which it is
firmly fastened. Nails grow from the root; and, as

---

[1] This idea of having one's hair " stand on end " passed long ago into common speech and classic literature.   Compare Job iv. 15.

[2] The total number of hairs on an average head is estimated at about 120,000.
The common diameter of an average hair is about $\frac{1}{40}$ of an inch.   The strength of
the hair is much greater than one would suppose.  A single hair has held a weight
of nearly $2\frac{1}{2}$ ounces without breaking.

long as this is not injured, they are not lost or disfigured by slivers, sores, blows, and bruises.

Disease or injury of the root generally results in a badly-shaped nail. The nails serve by their horny texture to protect the outer portions of the ends of the fingers and toes from injury, and to give a support for the fleshy ends.

Our finger-nails grow out about three times a year. They should be trimmed with the scissors once a week, leaving them long enough to protect the ends of the fingers. Nails should never be trimmed to the quick. They should not be cleaned with any thing harder than a brush or bit of soft wood. They should not be scraped with a penknife or anything metallic, as it destroys the delicacy of their structure, and will give them an unnatural thickness.

"Hang-nails" are caused by the skin sticking to the nail, which, growing outward, drags the skin along with it, stretching it until one end gives way. To prevent this, the skin should be loosened from the nail once a week, not with a knife or scissors, but with something blunt, such as the small end of a tooth-brush, or the ivory instrument made for the purpose.

**139. The Oil-Glands.** — The **oil-glands** are little round sacs, clustered together like a bunch of grapes, with a tube which opens into the hair-follicles.

Generally there are two to each hair; but in some places there are as many as four to eight around a hair, making a kind of collar about it. These glands furnish a natural dressing for the hair, and keep it moist and glossy. They also keep the surface of the skin soft and flexible, and prevent it from becoming dry and hard. In some places these glands, as upon the nose, chin, and forehead, are large, and the hairs very small; hence it often occurs that they open directly upon the skin. In these the oil is likely to collect, and get hard.

Bits of dust get into these glands, acting like plugs, and show themselves as small black specks, incorrectly called "flesh-worms," because of the resemblance which these little masses have to a worm. This oily secretion, which might well be called nature's hair-oil, is perfectly fluid in a healthy skin, and at the temperature of the body.

**140. The Sweat-Glands.** — The sweat-glands consist of very fine tubes, about one-tenth of an inch long, coiled up into knots, from each of which a canal, called a sweat-duct, opens on the surface of the skin.

The openings of these sweat-glands are arranged somewhat regularly, as may be seen by a common magnifying-glass, especially on the palms of the hands, between the ridges of the skin. On the sole of the foot and the palm of the hand they are very numerous, there being some three thousand of them to the square inch; while on the cheeks there are only five hundred and fifty in the same space, and twelve hundred to the square inch on the forehead. There are about three millions of them, it is said, in the whole body; and, if they were laid end to end, they would stretch to a distance of three to four miles.

These glands secrete the sweat, which is a colorless fluid, with a peculiar odor. It is a part of the waste matter of the tissues, which has been filtered from the blood, and is got rid of through these busy little glands in the skin. They are always at work pouring out sweat, though it may not be evident to the eye or touch. In hot weather, or after violent exercise, it is much increased, and collects in big drops, which run away in streams. The average daily quantity of sweat is not far from two pints. It varies very much according to what we are doing, the condition of health, how we are clothed, and the temperature of the surrounding atmosphere.

The object of this sweating through the skin is to regulate the temperature of the body by evaporation from the surface. We fan ourselves on a hot day to hasten this evaporation of the moisture on the skin. In hot weather, after taking hot drink or a hot-air bath, the skin does its best to reduce the temperature, and thus works all the harder in pouring out the sweat more profusely. When one is sweaty, it is highly imprudent to sit in a cool draught; for this evaporation may be suddenly checked, and we are apt to take cold.

It is now easy to see, that if this vast secreting-surface is hindered in its action by chilling it in a draught, by being too thinly clad, or by sudden changes in the weather, we may readily "catch cold," and cause a "congestion," as it is called, of certain internal organs, like the lungs, kidneys, or intestines.

141. **Why we should take Care of the Skin.** — There are nearly three millions of sweat-glands in the skin, acting like drainage-tubes, nearly three miles long, together with the oil-glands, pouring out two pounds

daily of sweat, oil, and other used-up matters, through the hard-worked skin. The sweat evaporates, and leaves the solid and oily matters to plug the mouths of these tiny sewer-pipes.

The dead scales of the scarf-skin are continually dropping off. They become sticky with the oil, and, getting entangled in the meshes of the clothing, become glued in a kind of thin crust to the surface of the body. This, if not regularly washed off, attracts the floating dirt and dust. Thus, the glands of the skin get choked up, and are not able to do their work properly.

Other organs, such as the lungs and kidneys, have to do their own work, and that of the skin besides. The balance of health is disturbed, because the blood is not properly purified; and disorders, especially of the skin, are almost sure to result.

Hence we see both the importance and the real need of keeping our bodies clean and neat. The fireman is ever busy rubbing and polishing his engine, to rid it of the dust, oil, and dirt, and to make it bright, clean, and neat. He is thus able to have all the parts of the machinery running smoothly and without friction. Surely we should take as good care of our person, especially of its surface, which is always sifting out impurities, and getting clogged with dust and dirt.

Finally, a filthy skin often becomes a breeding-place for the spread of those diseases and ailments which are "catching," such as scarlet-fever and measles, to say nothing of certain contagious disorders of the skin, by no means dangerous, but decidedly unpleasant.

**142. Baths, how and when to take Them.** — The first object in using soap and water on the skin is to keep it clean ; the second, to give vigor and strength to the whole body.

It takes very little time, expense, or bother to take a daily bath of some sort. A hand-basin, a sponge, a strip of cotton-flannel, a piece of castile soap, a gallon of water, and a towel, are all that are required. Even rubbing the body every day, first with a damp towel, and afterwards very briskly with a dry one, will, in most cases, keep the skin clean enough during the week, and give a healthy reaction, provided a bath with warm water and soap is taken at the end of the week.

Whether the daily bath should be taken in warm or cold water depends upon circumstances. Most persons, especially the young and vigorous, soon get used to cool, and even cold, water baths. The point is, to get a brisk and rapid reaction before the "shivers" come on. If we shiver after a bath, instead of getting a warm, comfortable glow, warmer water should be used.

The first effect of any cold bath is to shrivel up all the vessels of the skin, and make the surface pallid. Brisk rubbing should soon bring on a reaction, as it is called, in which the skin becomes red and full of blood. A feeling of genial warmth is felt all over the person. Always stop bathing if the shivering comes on, and use the towel vigorously until reaction sets in.

Young children and old people, unless strong, vigorous, and well used to it, cannot take a cold bath without some risk. Like all other things, it may be weakening to carry bathing to excess. Very much depends upon one's occupation and the condition of the skin. In some work, the grime, dust, and sweat must be washed off at least once a day, to feel at all comfortable, to say nothing of the health.

Coarse and rough towels should always be used if the skin will bear it. Some skins are very active common sewers, which are ever sending out a large amount of waste matters. In such cases, a daily bath, especially in hot weather, is almost a necessity. Hot baths, with hot drinks, causing free sweating, helped on by wrapping the person snug in bed, with a jug of hot water or a hot flatiron at the side or feet will often save children and others from illness, if promptly and vigorously done after exposure to cold or wet.

Swimming in running fresh water or in the sea has a wholesome effect on the skin, and is one of the healthiest of all exercises. Young people of both sexes should be taught to practise it whenever it is convenient. Never go in swimming when the body is overheated or very tired. Better sit down quietly, and cool off for half an hour.

Many are drowned every year from ignorance or carelessness in this one matter alone. The risk, of course, is from sudden cramps and colicky pains, which cause even a strong swimmer to sink like a lump of lead. For the same reason, it is not safe to take a swim just after a full meal.

It is a good time for vigorous persons to take a cold bath just after getting out of bed in the morning. A bath at bedtime is refreshing, and favors sound sleep. There is little risk of taking cold if we go to bed at once. Young and feeble children should bathe two or three hours after breakfast.

It adds to our health and comfort to keep the hair clean. The oil-glands get clogged; and dust and dirt rapidly make a coating on the scalp, and get entangled in the hair. Hence the hair should be washed, combed, and brushed, often and well. An occasional shampoo at home, with a wash made of the white of an egg and soapsuds, is healthful. Even a little borax dissolved in plain water, with vigorous rubbing, will do much to keep the scalp neat and clean.

143. **Why we need Clothing**. — Clothes serve to keep up an even temperature about the surface of the body. In winter, they keep in the bodily heat, and protect us from cold. In summer, they shield us from the direct rays of the sun, and from excessive heat. The temperature of our bodies varies only a few degrees at the most from 98½° F. In our climate, the outside air has a yearly range from 100° or more in midsummer, to 20° or more below zero in winter. We may even have a daily change of 20° to 40° F. The body is clearly much warmer than the surrounding atmosphere, and so it is continually parting with heat to the air.

These frequent changes are a severe tax on the body.[1] Without the protection of clothing, we could not endure the strain on the system. In short, clothing is our chief weapon of defence against the frequent changes of weather in our fitful climate.

[1] We often fail to recognize the danger of exposure to cold. One week's continuous cold in London, it is said, adds over two thousand deaths to the usual average; and they mostly occur among the poor, under-fed, and badly-clothed class of the population.

**144. Hints on the Use of Clothing.** — Clothes should be changed according to the climate or time of year. It is not prudent to leave off winter clothing too early in the spring, for our seasons are most uncertain. Woollen should be worn next to the skin, whether in summer or winter. A most imprudent but common error of our daily living is to take off our winter flannels some day in early summer because it happens to be warm. With our sudden changes of temperature, a person may run great risk of taking severe colds, pneumonia, and even "quick consumption." We should never allow ourselves to feel cold. If we cannot go where it is warm, or put on warm clothing, we should exercise until we feel warm.[1]

To keep our persons neat and clean, we must change our clothes often. This not only applies to garments used for daily wear, but to bed-clothes and night-clothes. No one should sleep in the clothes he wears during the day.

Under-garments should be frequently and regularly changed. All bed-clothes should be exposed freely to the light and air. Young children are less able to resist cold and sudden changes than grown-up people, hence great care must be taken in clothing them.[2]

The legs and chests of children should not be unduly

---

[1] "We should put off our winter clothing on a midsummer's day, and put it on again the day after. Only fools and beggars suffer from the cold; the latter not being able to get sufficient clothes, the others not having the sense to wear them." — BOERHAAVE.

[2] The celebrated surgeon, John Hunter, gave three simple rules for the rearing of children; and they are just as applicable to adults. They are, "plenty of sleep," "plenty of milk," and "plenty of flannel."

. exposed to the bitter blasts of winter, nor the cold east winds of spring.   Hundreds of children die every year from lung-diseases, due to ignorance or neglect in this matter.   Never wear wet or damp clothes one moment longer than necessary.   If you have on wet clothes, take the shortest way home, rub down thoroughly, and put on at once dry, warm clothes.

Do not let your damp skirts, wet stockings or shoes, dry on you, but always change them at once if possible. Neglect of this precaution is a fruitful cause of rheumatism and chest-ailments, especially among delicate people and young women.

Do not wear the clothing too tight, and thus allow it to interfere with free movements and easy and graceful carriage, to say nothing of health.   Garters and elastics worn below the knee are apt to hinder the circulation, and cause cold feet, and sometimes enlarged veins.

Dresses and skirts should never drag their full weight from the hips, but should be supported from the shoulders.   Health and comfort should not be sacrificed for any strange devices to dress the person in a slavish submission to any fashion.

Teachers and parents should take the utmost care in the matter of children's clothing.   See to it that they have on the proper outside garments on going out, and that they take them off on coming in-doors.

Children, and older people too, should never run out-doors without proper covering for the head.   Never allow pupils to sit in the schoolroom with outside garments on, such as waterproofs, gossamers, cloaks, rubbers, rubber boots, and leggings.

**145. Effects of Alcoholic Drinks and Tobacco upon · the Skin.** — Now, it is very evident that the skin, like any other active organ, depends for its nourishment upon the proper circulation of the blood. Hence, if this is interfered with by alcoholic drinks, the skin lacks its chief element of vitality.

The effect upon the skin is soon evident. At first the skin has a peculiar soft, satiny feeling, from which sagacious physicians have discovered the alcoholic habits of their patients, and perspiration is easily induced. Later on, the skin becomes thick and discolored, sometimes red and sallow. Dark-brown blotches begin to appear.

Again, the skin is often affected with inflammations of various sorts, leading to many different kinds of unseemly eruptions, boils, and abscesses. Alcohol often aggravates a skin disease, and tends to keep up the inflammation of any existing skin disorder, as salt-rheum, nettle-rash, and other ailments.

The blood-vessels of the skin on the face, especially the nose and cheeks, become permanently stretched. The red and swollen nose, known as the "bottle-nose," due to alcoholic drink, is an every-day sight on the street.[1]

The effect of alcohol upon the bodily heat, in which the skin plays so important a part, has been discussed in chapter ix.

The pores of the skin of a tobacco-user become saturated with the poisonous nicotine, giving the skin a peculiar dry and sallow look.

---

[1] "Frequent repetition of alcoholic drinks tends to permanently impair the activity of the circulation of the blood in the skin. Hence the visible vascular twigs and rubicund nose that characterize the physiognomy of the habitual drinker."—DR. JAMES C. WILSON.

# CHAPTER XI

## THE NERVOUS SYSTEM

**146 How all Parts of the Body work together in Harmony**. — We have studied the human body as a kind of living machine. We have examined its various parts, and found them adapted to some special work essential to the well-being of the whole.

Each organ not only looks after itself, but it is ever ready to come to the help of other parts of the body. Everywhere we find organs working together for each other's good. Strike suddenly at the eye, and the lids fall to protect it. Tickle the foot, and the muscles of the leg contract, and pull it away. When the skin is inactive, the kidneys come to its help. Fifty skilled mechanics might do their best at building a vessel or a house; but if each man worked as he pleased, and took no heed of the rest, the result of their work would be of little account. The master-builder must be at his post, skilful to direct, and quick to act.

So it is with our bodies. The wonderful agency which directs and governs every organ of the body is the nervous system.

**147. The Nervous System**. — The **nervous system** is to the organs of the body what steam is to the engine. Shut off the steam, and the rods, wheels, and bands, which a moment before were whirling round, at once become still and useless. So with the body : any injury to the nervous system may paralyze some or all of the various organs, and at last produce death.

The nervous system may be compared to nothing so aptly as to a complete telegraphic system. The brain

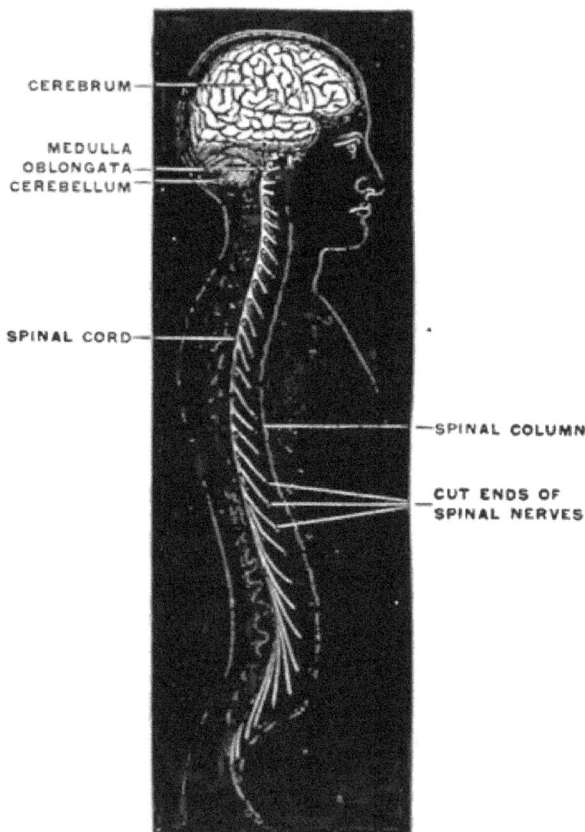

FIG. 59. — Diagram of Brain and Spinal Cord.

itself is the main office ; and the thousands of nerve-fibres, branching off to all parts of the body, are the telegraph-wires. Telegraphic despatches are constantly being sent to the brain, to inform it of what is going on in various parts of the body. The brain, on receiving the news, at once sends back its commands as to what must be done.

Thus, if we accidentally pick up a hot coal, we drop it instantly. A message is sent from the nerves of touch in the fingers to the brain ; and, when the brain learns that the fingers are being burned, it hurries off its orders to the muscles of the fingers to drop the burning coal. We might multiply these examples to almost any extent. Thus every effort of the will we make originates in the brain, and is carried out under the direction of the nervous system.

' Now, you must know that the pain is not really in any organ, but in the mind itself, because this alone can feel. If we strike the " funny bone " (the place where the nerve is in an exposed place as it runs over the elbow to the fingers), we feel pain in the third and fourth fingers. The brain refers the pain to the end of the nerve. Thus, after a limb has been amputated, any injury to the stump will be referred to the place where the nerve once led.[1]

The nervous system consists of two distinct parts : —

1. The **brain, spinal cord,** and **cerebro-spinal nerves,** forming an unbroken connection between the brain and all parts of the body.

---

[1] " When the current of a battery is applied to the nerves of an arm-stump, the irritation is carried to the brain, and referred to all the regions of the lost limb. On one occasion a man's shoulder was thus electrized three inches above the point where the limb was cut off. For two years he had ceased to be conscious of his limb. As the current passed through, the man, ignorant of its possible effects, started up, crying, ' Oh, the hand ! the hand ! ' and tried to seize it with the living grasp of the sound fingers. No resurrection of the dead could have been more startling." — DR. S. WEIR MITCHELL.

2. The **sympathetic system,** which is connected mainly with the organs of digestion, circulation, and respiration.

**148. Nerve Tissue.** — Nerve-tissue is the soft, marrow-like substance which forms the principal bulk of the brain, the spinal cord, and the nerves. In the brain the inner part is white, and the outer layer is a gray or grayish-pink substance. In the spinal cord the inner part is gray, and the outer white. The gray is richer in blood-vessels than the white.

All muscular action, we have learned, results in the oxidation, or burning, of the muscles themselves. In the same way the work of the brain causes the oxidation, or burning up, of its nerve-tissue; so that every thought, every sensation, every effort of the will, which goes from the brain, destroys a part of its substance.

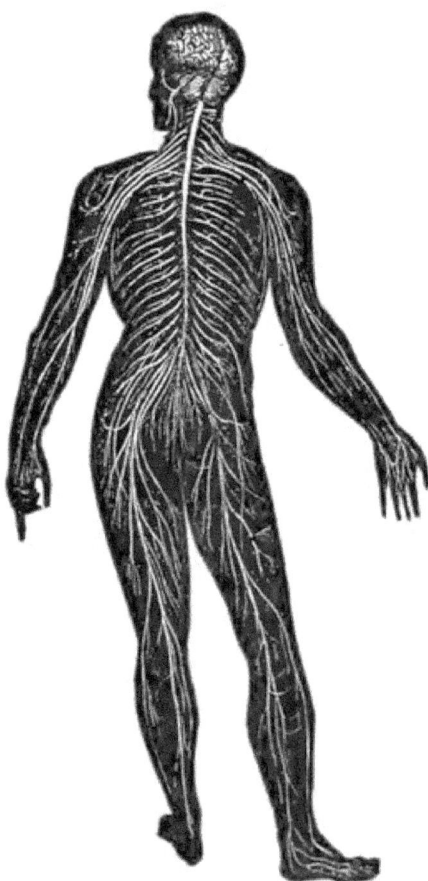

Fig. 60. — Diagrammatic view of the Brain, Spinal Marrow, and Nerves (from behind).

The **nerves** are really so many portions of the brain and spinal cord, extending into every minute part of the body. A nerve is a slender silvery-white cord when seen by the naked eye. Under the microscope, this white part is seen to consist of bundles of delicate little fibres.

**149. The Brain.** — The brain is the organ of the mind; in other words, it is the seat of the consciousness, the intellect, the memory, the will, the affections, and the emotions.

This important organ fills the entire cavity of the skull, and consists of a number of separate masses of nerve-matter abundantly supplied with blood-vessels. Each separate mass is the main workshop for some special department of the work of the nervous system, and has little or no connection with the other parts.

The average weight of the brain is about fifty ounces, or about three pounds.[1] A few cases have been noted in men of great mental capacity, in which the brain weighed sixty-four ounces. Such instances are, however, far from common. As a rule, a large brain is the sign of a vigorous mind and superior faculties; and a healthy brain, of a sound, healthy mind.

The three principal masses which compose the brain are, (1) **the cerebrum, or brain proper;** (2) **the cerebellum, or lesser brain;** (3) **the medulla oblongata.**

---

[1] Daniel Webster's brain weighed fifty-three and a half ounces, and Ruloff's — a notorious murderer, but in some respects a very learned man — fifty-nine ounces. The brain of Cuvier, the celebrated naturalist, weighed sixty-four and a third ounces; and that of Dupuytren, a famous French surgeon, sixty-two and a half ounces. The hats of ten gentlemen were tried upon the skull of Robert Burns, and the only one of the ten that could cover it was the hat of Thomas Carlyle. An idiot's brain rarely exceeds thirty ounces.

150. **The Cerebrum, or Brain Proper.** — The **cerebrum** fills the whole of the upper part of the skull, and is nearly seven-eighths of the entire mass. It consists of two parts, or halves, almost entirely separated from each other by a deep cleft, or fissure, from front to

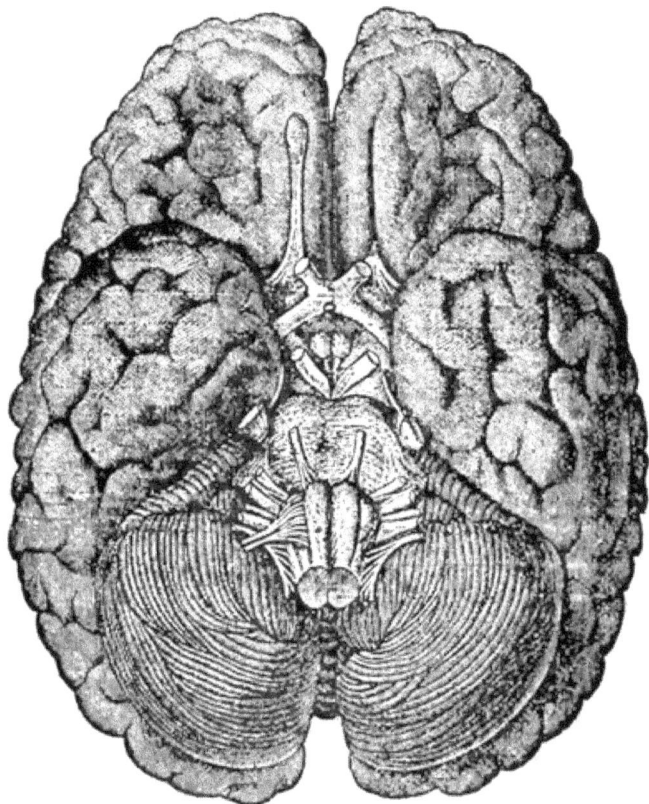

FIG. 61. — Under Surface of the Brain.

back. Each of these halves — or hemispheres, as they are called — consists of three portions, or lobes, so that the cerebrum is made up of six distinct parts.

The cerebrum has a peculiar folded-up appearance; its various folds, or convolutions as they are called, being separated by deep clefts, sometimes nearly an inch deep. In this simple way the surface of the brain is increased many fold. The interior part of the brain is made up of the white nerve-substance just spoken of. The gray matter is the outer layer, about one-eighth of an inch in thickness, and is spread over the white substance like a handkerchief crumpled up.[1]

The active powers of the brain are supposed to reside in this outer layer; and these powers are great or small, according to the number and the extent of the folds, or convolutions. In the lower animals the brain has no folds; but as we pass to animals of a higher grade, the folds begin to appear.

**151. The Cerebellum, or Lesser Brain.** — The cerebellum, or little brain, lies beneath the back part of the brain proper, from which it is separated by a fold of the dura mater. It is made up of two halves, each formed of a number of layers of gray and white nerve-matter, curiously arranged, resembling somewhat the branches of a tiny tree.

The functions of the cerebellum are not yet certainly known. It is supposed to exercise an influence over the muscles of the body, in regulating their movements.

---

[1] The brain is enclosed within three distinct layers, called its membranes, — the *dura mater* ("hard mother"), the *arachnoid* (like a spider's web), and the *pia mater* ("delicate mother").

The dura mater is the tough membrane which lines the inner surface of the skull, and forms a loose outer covering for the brain. The middle layer, called the arachnoid, secretes a fluid which keeps the inner surface moist. The *pia mater* is a very delicate membrane which dips down between and lines the folds of the cerebrum, rich in blood-vessels which nourish the brain.

**152. The Medulla Oblongata.** — The medulla oblongata is the thick upper part of the spinal cord which is held within the cavity of the skull. It is just under the little brain, and makes the connecting link between the brain and the spinal cord. It is a small affair, only an inch and a half long ; but it is a highly important part of the brain, since from it arise the nerves which regulate breathing, swallowing, the heart's action, and so on.

The seat of sensation is believed to be lodged in the upper part of the medulla oblongata. It has also control over the action of the lungs and the heart. If this part of the brain be broken or cut, respiration and circulation will at once cease, causing instant death.

**153. The Cranial Nerves.** — From the brain proceed twelve pairs of nerves, called cranial nerves. They pass out of the skull through little holes in its base, and supply the face, certain internal organs, and the organs of smell, taste, hearing, and sight. These cranial nerves are of three kinds, — sensory, motor, and mixed ; i.e., combining both.

The tenth pair is called the "wandering nerve." [1] This is perhaps the most important nerve in the body. It supplies the larynx, the lungs, the heart, the stomach, and the liver. It is partly motor and partly sensory.

**154. The Spinal Cord.** — The spinal cord, or marrow, is a cylinder of soft nerve-tissue, extending from the base of the skull to the region of the loins, where it tapers into little threads. It is a continuation of the

---

[1] Pneumogastric.

medulla oblongata, and its average length is about eighteen inches.

Like the brain, the spinal cord consists of the two
kinds of nerve-matter, — white and gray : but their position is reversed ; the gray being in the inside, and the
white outside.

The spinal cord receives impressions from various
parts of the body by means of its sensory nerves, and

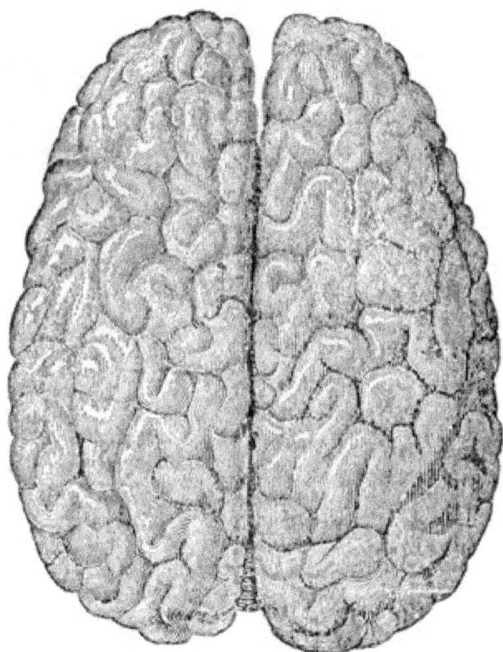

FIG. 62. — Upper Surface of Brain, showing the Convolutions and its Double Structure.

carries them to the brain, where they excite sensation,
or consciousness.

Again, it sends out, by means of its motor nerves,
the commands of the brain to the voluntary muscles.

**155. Reflex Action of the Cord and Brain.** — The spinal cord is not merely a bundle of nerve-fibres for carrying messages to and from the brain. It has a certain power of its own. It acts as a kind of independent centre, receiving messages, or sensations, from certain parts of the body by means of its sensory nerves, and on its own authority sending back orders to the muscles

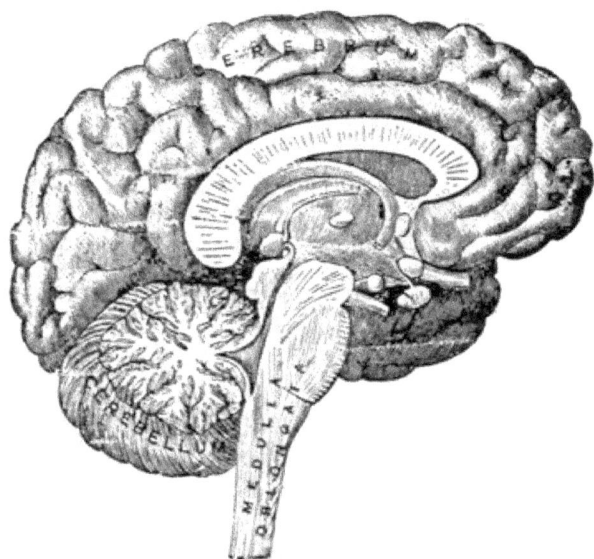

FIG. 63. — Vertical Section of the Brain.

by its motor nerves, without waiting to consult the brain.

This is what it known as **reflex action.**

If one is asleep, and the feet are gently tickled, the legs will be moved out of the way without the sleeper necessarily being awakened. When the spine is broken by an injury, causing pressure upon the cord, then all sensation and motion are lost in the paralyzed limbs. But if these paralyzed limbs are irritated, as by pricking the soles of the feet, then the legs kick out vigorously. The injured person does not feel the pricking, and can exercise no control over the kicking legs. There is here no conscious action whatever.

This unconscious action is the result of reflex action of the spinal cord. It is called reflex because the impression does not go to the brain, but is "reflected," meaning "to turn back again," to the seat of injury, from the sensory nerves through the motor nerves.

156. **Importance of Reflex Action.**— We rarely stop to think how important reflex action is to our health, comfort, and safety. Because we are able to do hun-

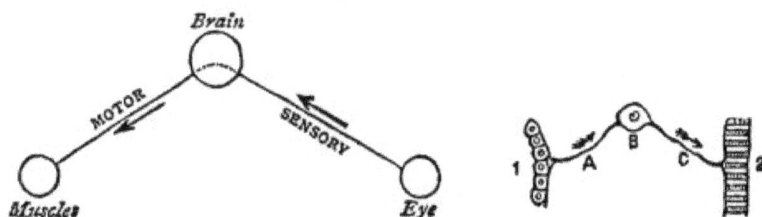

FIG. 64. — Diagrams showing the mechanism of a simple reflex action: 1, surface, say of mucous membrane; 2, muscle; A, sensory nerve; B, reflex centre; C, motor nerve.

dreds of things every day without any effort of the will, we are apt to forget its importance. In fact, the greater part of nerve-power expended in the body goes to produce these numberless reflex actions. We are not so independent as one would at first think. Ten thousand acts take place which tend to govern and preserve our health. We have as little control over them as we have over the stars above us.

We have already been told of a few reflex acts. Let us call to mind a few more familiar illustrations. If our feet slip on the ice, without the effort of the will the body tends to recover itself. The mind does not always act, at least in the ordinary way, to pull the fingers away when they touch a hot stove. We try to brush the flies away when we are asleep. By an effort of the will, we can stop our breath for a moment or two : but soon the call for air is imperative ; and the order must be obeyed, whether we will or no. The great work of digestion is going on day after day, but we have no control over its complicated movements.

By this wonderful provision of nature the " think-centres " are relieved of a vast amount of work. If we were forced to use our will-power at every step in the process of digestion, the brain would be put to a severe strain. We could not eat, and then quietly go about our business. If we had to plan and will every heart-beat, we should soon be ready to give up the struggle for life.

If we had to exert our will every time we breathed in or out, we would soon get tired of it, and long to die. We could never sleep ; for the brain would have to be on the alert to decide if it were time for the next heart-beat, the next inspiration, and the proper time for each digestive fluid to flow.

**157. The Spinal Nerves**. — From the spinal cord thirty-one pairs of **spinal nerves** proceed to the trunk and limbs.

They pass out on each side of the spinal canal through small holes in the sides of the backbone. Each of these spinal nerves has two roots, — one going from the front part, and the other from the back part, of the cord. These two roots unite, and run side by side, forming one silvery thread as they pass out from the backbone.

The root which goes from the front, or anterior, part of the spinal cord, is the **motor** nerve, and controls the muscles of the body.

The one which comes from the back, or posterior, part of the cord, is the **sensory** nerve, and carries sensations from the various parts of the body to the spinal cord.

As each nerve-trunk leaves the backbone, it subdivides, and sends off branches into various parts of the body. If any one of these nerves is cut or injured as it leaves the backbone, the power of feeling and movement ceases in all those parts to which it is distributed ; that is, those parts of the body are paralyzed.[1] The case is practically the same as cutting a telegraph-wire, and thus stopping the passage of the electric current. To a certain extent this is the case when our leg or arm " goes to sleep." The fact is, some of the nerves have been pressed upon, and their action is stopped for the time being. Remove the pressure, and the nerves regain their power.[2]

**158. The Sympathetic System of Nerves.** — Besides the system of nerves just described, there is another

[1] A very important and interesting point to notice, is that, as the motor nerve-fibres leave the brain, they cross over from one side of the spinal cord to the other. The sensory nerve-fibres cross in the same way soon after entering it. Thus, the right half of the brain governs the left half of the body, and the left half of the brain controls the right side of the body. A blow on either side of the brain may therefore produce paralysis of the opposite side of the body. Thus, if we see a person with left side paralyzed, we know that the seat of injury is on the right side of the brain.

[2] Imagine a telegraph-cable of delicate India-rubber tubes filled with mercury ; a squeeze would interrupt the continuity of the cable without destroying its physical continuity. Remove the pressure, and the current would pass along the cable as usual.

set of nerves, known as the **sympathetic nervous system.**[1]
It consists of a double chain of nerve-knots, or ganglia, connected by nervous cords running down in front
and on each side of the backbone. This chain of
nerves is not shut up within the bony tube formed
by the skull and spine. The knots of nerves are connected with each other, and with the sensory roots of
the spinal nerves, by a network of gray nerve-tissue.

From these knots, nerves go to all the internal
viscera, forming a complete system by themselves, and

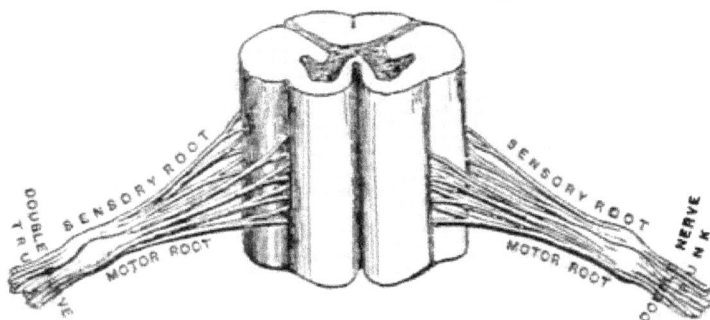

Fig. 65. — Section of Spinal Cord, with Roots of Spinal Nerves. Front view.

acting almost independently of the cerebro-spinal system. A close network of the sympathetic nerves is
spread round the muscles of the heart, the lungs, the
stomach, and the intestines, as well as round the walls
of the minute arteries and capillaries.

Compared with the cerebro-spinal set of nerves, the
sympathetic nerves are very slow to act. A blush
steals slowly up to the roots of the hair. If we go
from the dark into a strong light, we are blinded; the

[1] The name "sympathetic" is given to this part of the nervous system,
because it was thought that, through its agency, distant organs have sympathy
with one another's afflictions. Thus, for example, severe pain in any part of
the body will cause some sensitive persons to be sick at the stomach.

pupil is too large to endure it. An impression is made
on the sympathetic
nerves, which cause
the pupil to contract
slowly while we are
shading our eyes.

**159. The Health
of the Nervous Sys-
tem.** — It is not nec-
essary to discuss in
detail the health of
the nervous system,
as we have done with
the other parts of the
body. The health of
every tissue and of
every organ of the
body is dependent
upon, and woven in-
to, the welfare of the
nervous system.[1]

[1] "As a king sits high above
his subjects upon his throne,
and from it speaks behests that
all obey, so from the throne of
the brain-cells is all the kingdom
of a man directed, controlled,
and influenced. For this occu-
pant, the eyes watch, the ears
hear, the tongue tastes, the nostrils smell, the skin feels. For
it, language is exhausted of its
treasures, and life of its experience; locomotion is accom-

FIG. 66. — The Great Sympathetic Nerve.

plished, and quiet insured. When it wills, body and spirit are goaded like over-
driven horses. When it allows, rest and sleep may come for recuperation. In
short, the slightest penetration may not fail to perceive that all other parts obey
this part, and are but ministers to its necessities." — *Odd Hours of a Physician.*

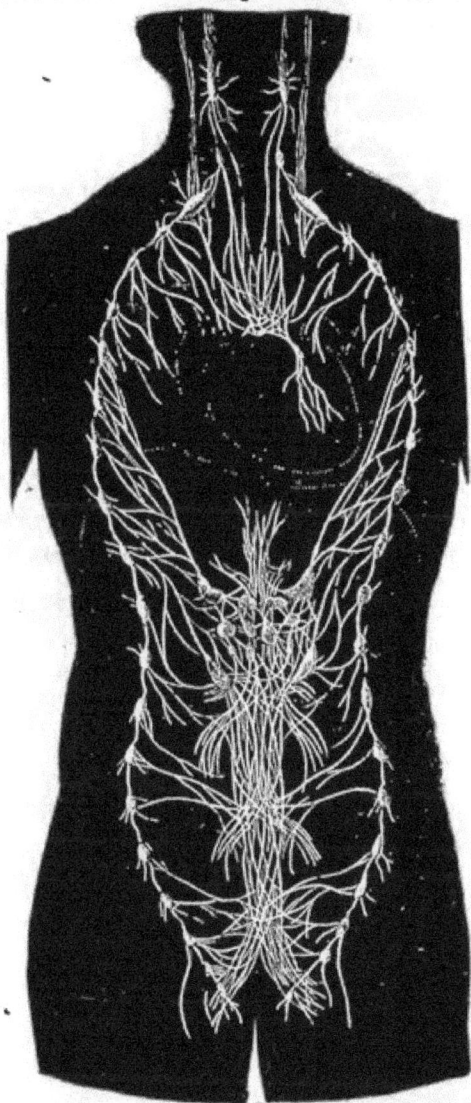

A blow on the head often causes vomiting.  If a tiny blood-vessel in the brain is broken, and forms a little clot as large as a pea, a paralysis of one side may be produced by its pressure on the delicate brain-tissue. An overloaded stomach may make the brain dull and stupid for some time.  Ill-digestion may change the disposition, and make one cross, morose, and ugly The loss of sleep may cause exquisite suffering.

In early English history, people who were con-demned to death, by being prevented from sleeping, always died raving maniacs.  Those who are starved to death become insane.  The brain is not nourished, and they cannot sleep.  A blow on the head, and cer-tain diseases, may produce a profound and mysterious effect upon the nervous system.

On the other hand, severe accidents to the brain may not produce serious results.[1]

Like any other organ, the brain may be strengthened and increased in its power by use and education.  Im-pressions made upon the mind in early life are more readily received and more completely retained than those which are made when the growth of the brain is far advanced.  For this reason, education should be

[1] A pointed iron bar, three and a half feet long and one inch and a quarter in diameter, was driven by the premature blasting of a rock completely through the side of the head of a man who was present.  It entered below the temple, and made its exit at the top of the forehead, just about the middle line.  The man was at first stunned, and lay in a delirious, semi-stupefied state for about three weeks.  At the end of sixteen months, however, he was in perfect health, with the wounds healed and with the mental and bodily functions unimpaired, except that the sight was lost in the eye of the injured side.  He died nearly thirteen years after the accident.  The iron bar and the skull may be seen at the Museum of the Harvard Medical College, in Boston.

begun early in life. It is an object for which most parents are willing to work hard, and exercise much self-denial, to accomplish.

**160. Abuse of the Nervous System.** — Just as the stomach may be overworked, and fail after a time to digest food properly ; and as muscles are exhausted by over-exertion, — so may the nervous system, especially the brain, be overtaxed.

Mental work is rarely hurtful to a healthy person who takes good care of himself. It is not so much the severe mental toil, as it is worry, that makes the mental wreck. It is not study, but fretting, that causes the student to break down in his studies. It is the wear and tear of the nervous system in the fierce struggle for wealth and position, that makes an old man out of a young one before he is forty. Let a child have plenty of nutritious food properly given, plenty of sound sleep, enough of suitable clothing, and a calm and wise oversight at home, and he will rarely be injured by too much study.

It is fretting about passing examinations, worrying about promotions, and other baneful influences which have become attached to our educational system like barnacles to a stately ship, that make the delicate, sensitive child cross, peevish, and sickly, and lay the foundation for years of ill-health.

Every part of the nervous system is busily at work doing its allotted duty. The vast sympathetic system, which lies at the very foundation of our nutrition, is severely taxed to do its work properly. Now, let a person fret and worry day after day over real or fancied troubles, abuse his digestive organs by too much or too little food, go without proper sleep, turn night into day, try to prop up his flagging energies with more or less

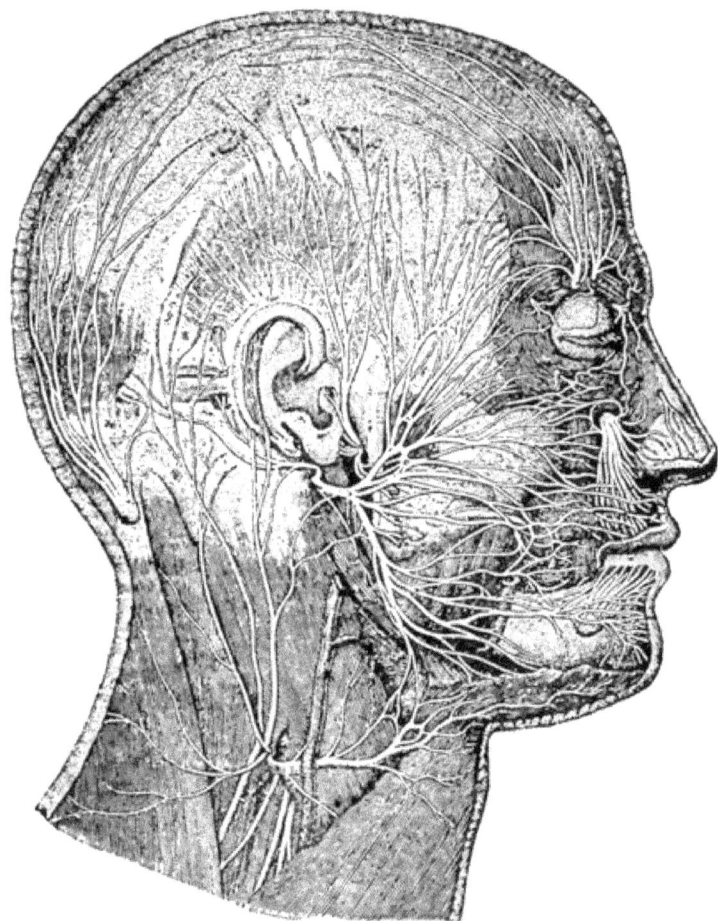

FIG. 67. —Superficial Nerves on the right side of the Neck and Head.

of alcoholic liquors, and the strain on the nervous system will make a physical and mental wreck.

Like a spendthrift, who lavishly spends his principal,

and persists in calling it his income ; so is a man, who is indulging in all these forms of dissipation, really exhausting the limited amount of nervous force at his command. Unhealthful and injurious habits, whether in the important or comparatively trifling matters of daily living, are drafts drawn on the future, which must be met at no distant day with all the attendant perils of physical or mental bankruptcy.

All forms of nervous gratification that pander to the lower tastes, as too much reading of sensational litera-ture, frequent attendance at low theatrical performances, and an over-indulgence in gross pleasures of every kind, cause a great waste of nervous tissue, and may in time sap the foundation of health and happiness and leave its victim at last a wreck of his former self.

**161. Sleep.** — Sleep is necessary to life and health. In our waking-hours, rest is obtained only at short in-tervals. The muscles and nerves, the brain in particu-lar, are in full activity when we are awake. Repair goes on every moment, whether we are awake or asleep; but during the waking-hours the waste of the tissues is in excess of the repair, while during sleep the repair exceeds the waste. Hence the good mother Nature, at regular intervals, causes all parts of the bodily machin-ery to be run at their lowest point. In other words, we are put to sleep.[1]

To be sure, the heart beats, the lungs take in air, and the stomach digests its food ; but these great

---

[1] The great classic writers of all time have symbolized sleep under all kinds of fanciful expressions. Shakespeare alludes to sleep as " the golden dew," " our foster-nurse of nature," " nature's soft nurse," " death's counterfeit," " the death of each day's life," " great nature's second course," " chief nourisher in life's feast."

organic processes are carried on at their lowest point. The vital organs rest because they are worked at the lowest rate that will keep us alive. The eye, the ear, the brain, and the nerves, are at rest by darkness, silence, and unconsciousness. The tired muscle regains its vigor, and the exhausted brain is refreshed.

Sleep is more or less sound according to circumstances. Fatigue, if not too great, aids it ; while idleness lessens it. Some kinds of food, as tea and coffee, may prevent it. Anxious thought and pain, and even anticipated pleasure, may prevent it. Just as sleep is sound, the body and mind are refreshed.

The best time for sleep is at night. The soundest and best sleep is obtained during its silence and darkness. People who are forced to work nights, and to sleep during the day, have a strained and wearied look, which is easily recognized by the trained eye.

The amount of sleep depends upon our· occupation and our temperament. Some require little sleep, while others need a great deal. Eight hours of sound sleep for a grown man or woman, and more for children and old people, is about the average amount required. Children naturally need more sleep, because their bodies need more rest during the period of growth. Hence the infant sleeps most of the time, if well, and properly cared for.

Children should always be put to bed early, and allowed to sleep in the morning until they awake of themselves. During hot weather, the active child should be undressed, bathed, and put to bed in the middle of the day for a good nap. Do not go to bed

with the brain excited or too active. Read some
pleasant book, laugh, talk, sing, take a brisk walk, or
otherwise indulge in a little recreation, for half an hour
or so before going to bed.

162. **Narcotics.** — One of the most fruitful sources
of injury to the nervous system is the use of alcoholic
drinks, tobacco, opium, and other narcotics. It is because of their deadening and paralyzing effect upon the
nerves that they are called **narcotics,** or nerve poisoners.

As we have already seen, a little of any alcoholic
drinks has the power to create an inclination to take
them again. Yielding once, they beget an itching desire to repeat the action. This unhealthy craving may
pass away, but oftener grows fierce by what it feeds on.
A man is no longer master of himself, but is the victim
of a relentless craving for what may sooner or later
destroy him. It is almost impossible for him to free
himself from the invisible chains of his baneful habit.

163. **General Effect of Alcohol upon the Nervous
System.** — The real substance of the nerves, the great
nerve-centres, and the brain, is made up of nerve-pulp,
the softest and most delicate matter in the body. A
rude touch will crush and destroy it. It is very rich in
blood-vessels, which bring blood to renew and nourish
it, and also to carry away the waste.

This ceaseless interchange between the blood and the
nerve-pulp maintains the vital action and power of the
nervous system. No other part of the living body is so
sensitive to the presence of alcohol as this delicate
nerve-pulp. It has much water in it, to keep it in a moist
and workable condition, but alcohol rapidly takes to

itself the water of the nerve-pulp, thus causing injurious changes in the nerve-substance itself in proportion to the amount of the narcotic taken.

Again, when alcohol is once taken up by the nerve-tissue, it does not easily escape. In other words, this unnatural agent is shut up and imprisoned in the nerve-structure, and is only got rid of by a gradual and slow process. For these two reasons, therefore, the nerve-tissue is peculiarly subject to the injurious effect of alcoholic liquors in whatever quantity taken.

164. Alcoholic Liquors and the Nervous System. — The first symptom showing that the nerves are disturbed by alcohol is the quickened action of the heart, and with it the dilatation of the blood-vessels. The face is flushed, and there is a warm glow all over the body, because the nerves which control the action of the heart, and those that regulate the size of the blood-vessels, are paralyzed by the alcohol.

The tiny blood-vessels of the brain are overcharged, the brain becomes more active, thoughts flow more rapidly, and the speech becomes more fluent ; but this activity is only an unhealthy excitement. The power of right thinking is diminished, and the fluent speech is usually lacking in good sense. The use of alcohol, even in what is called moderate quantities, in the lighter liquors, such as beer, wine, and cider, tends through its action on the brain, to weaken the judgment and to blunt the moral sense.

The nerve-pulp of the higher, or controlling, part of the brain is next brought within the grasp of this deadly alcoholic influence, and the faculties of the mind are still more impaired. Reason is off duty; and the lower, or animal, impulses begin to manifest themselves. First, the control of judgment and the will [1] disappears; and the emotional, the impulsive, and the purely instinctive part of our nature is laid bare. It is a fact of common observation that alcohol is one of the greatest criminal forces in the world, and as we come to understand its action on the brain we see why this is so.

If still more alcohol is taken, those actions of life which are under the direct control of the spinal cord become disturbed. The power over some of the muscles is lost, and the energy of the whole muscular system is lessened. The muscles of the lower lip and the legs are the first to feel this unnatural torpor. The speech is thick, and the gait uncertain. Trembling, faintness, and vomiting are more or less frequent.

In the last stage of all, the paralysis. of the nerve-centres and of the brain is carried to its full extent. All the inlets of the senses are closed, all consciousness and sensation are lost, and all power over the voluntary movements is gone. The heart still beats, and the blood circulates, and the breathing still goes on; but these are the sole remnants of vitality, the slender threads by which a hold is retained upon life.

165. The Final Result. — After a time, the continuous use of alcohol steadily weakens the self-control of its hapless victim, and at last makes him an utter slave to

---

[1] " Under the influence of alcohol, the weakening of the will becomes excessive. The extravagances, violences, and crimes committed in this state are innumerable." — DR. J. RIBOT.

his lower nature.[1]   The craving for ardent spirit becomes well-nigh irresistible.   Self-respect, honor, con-
science, everything, is sacrificed to get fresh fuel for the
alcoholic fire which is burning up its victim.   The road
is now straight which leads to some private "retreat,"
or even the insane-asylum.

The disease known as *delirium tremens*, meaning a
trembling madness, is a terrible instance of the effect
which alcoholic liquors may exercise over the nerve-
centres.   It is marked by muscular tremors, a low temperature, cold and clammy skin, persistent wakefulness,
muttering talk, then the wildest delirium, with all the
horrors of hideous delusions which imagination can
possibly conceive.[2]

These extreme instances of the breaking-up of the
nervous system are found in our insane-asylums, — men
and women who have lost every trace of humanity ;
hopeless, helpless, doomed to a living death until they
cease to be.

The insanity of a large number of the inmates of
nearly all insane-asylums is due to the use of alcoholic
liquors.   In each case the first glass was the first step
toward this fate.

**166.  The Inherited Craving for Alcohol.** — The craving for alcohol may be inherited by its victim's innocent

---

[1] "O, that men should put an enemy in their mouths, to steal away their
brains ! that we should, with joy, revel, pleasure, and applause, transform ourselves
into beasts ! " — *Othello*, Act ii. Scene 3.

[2] " The victim almost always apprehends some direful calamity ; he imagines
his bed to be covered with loathsome reptiles ; he sees the walls of his apartment
crowded with foul spectres ; and he imagines his friends and attendants to be fiends,
come to drag him down to a fiery abyss beneath." — W. B. CARPENTER, M.D.

children. Just as we may inherit from our parents mental and physical vigor or weakness, our features, and even moral traits, so it is possible to inherit a thirst for alcoholic drink. This inherited curse of strong drink has caused many a good family to "run out," and leave the children and grandchildren pitiable wrecks of humanity.

Again, the children of parents whose brains are injured by alcohol, are more prone than others to mental disorders, and to those widely varied diseases which, for lack of a better name, we vaguely call "nervous."

Weak moral natures, and a tendency to lying and deceit and other forms of wrong-doing, are only a few of many baneful defects which children may inherit from drinking-parents.[1]

**167. Tobacco.** — Tobacco, whether snuffed, chewed, or smoked, is a narcotic and a poison. Its injurious effects are due to its active principle called *nicotine*, which is of itself a narcotic poison.

Tobacco is hurtful to young people, and by no means free of harm to adults. It produces an artificial exhaustion, as it were, of the nerve-centres. The to-

---

[1] This point has been very tersely put by a sagacious physician, the late Dr. Willard Parker of New York. He says : "There is a marked tendency in nature to transmit all diseased conditions. Thus the children of consumptive parents are apt to be consumptive. But, of all agents, alcohol is the most potent in establishing a heredity that exhibits itself in the destruction of mind and body. There is not only a propensity transmitted, but an actual disease of the nervous system."

".The appetite for strong drink is frequently transmitted from parents to the children, just as other traits of the mind or body. Sometimes it develops early, sometimes late in life. As a rule this hereditary propensity shows itself at an early age, and is apt to be intensified at the age of maturity." — Dr. JAMES C. WILSON, *in Pepper's System of Medicine*, vol. v. p. 634.

bacco habit once acquired generally leads to continual
and increasing use.

Thus, after a time, tobacco produces functional derangement of the nervous system, palpitation of the
heart, certain forms of dyspepsia, and more or less
irritation of the throat and lungs.

Sometimes, after long smoking, a sudden sensation
of dizziness, with a momentary loss of consciousness,
is experienced.  At other times, if walking, there is a
sudden sensation of falling forward, or as if the feet
were touching cotton-wool.  While the stomach is
empty, protracted smoking will often produce a feeling
of nausea, accompanied with a headache.[1]

The brain will often become affected.  The ideas
lack clearness of outline.  The will-power may be
weakened, and it may be an effort to do the routine
duties of every-day life.  The old tobacco-user is often
cross, irritable, and liable to outbursts of passion.  The
memory is also quite often impaired for the same reason.  The narcotic principle, the deadly nicotine, has
become soaked into the delicate nerve-pulp retarding
its nutrition.  The nerve-centres are no longer able to
hoard up their usual amount of vital energy.  Hence
arise the many and various nervous symptoms due to
the poisonous effect of tobacco.

---

[1] The external application of tobacco to chafed surfaces, and even to the
healthy skin, will occasion severe, and sometimes fatal, results.  A tea made of
tobacco, and applied to the skin, has caused death in three hours.  A tobacco
enema has resulted fatally within a few minutes.  Smoking a large quantity of
tobacco at one time has been known to produce violent, and even fatal, effects.
Nicotine is one of the most rapidly fatal poisons known.  It rivals prussic acid
in this respect.  It takes about one minute for a single drop of nicotine to kill a
full-grown cat.  A single drop has killed a rabbit in three minutes.

**168. Effects of Tobacco upon Young People.** — Tobacco, in any form, has a peculiarly injurious effect upon young and growing persons. It not only stunts their growth, but produces a weakened state of the system, which tends greatly to impair muscular and mental activity. The profound effect that tobacco has upon the nervous system, after the first trial of smoking or chewing, is matter of familiar experience. Even after the system gets used to the narcotic, young people continue to suffer oftentimes from nausea, dizziness, headache, muscular trembling, loss of appetite, and general weakness.

**Evil Effects of Tobacco.** — " Tobacco reduces the intellectual power of a boy. It does this either by opposing mental application and effort, or else by producing deterioration of the intellect, probably both to a greater or less degree. — DR. E. O. OTIS, *in Boston Medical and Surgical Journal.*

" It is our deliberate opinion that the unsatisfactory recitations and consequent failures at final examination, so injurious to the interests of this establishment, are to be attributed, in great measure, to nervous derangement caused by the common use of tobacco by the students. It becomes our duty to recommend some stringent measures to correct this practice." — *Medical Report on the Use of Tobacco by the Cadets at the U. S. Naval Academy.*

The evils of tobacco are intensified a hundred fold upon the young. Here it is unqualifiedly and uniformly injurious. It stunts the growth, poisons the heart, impairs the mental powers, and cripples the individual in every way. Not that it does all this to every youth, but it may be safely asserted that no boy of twelve or fourteen can begin the practice of smoking without becoming physically or mentally injured by the time he is twenty-one.

"Sewer-gas is bad enough, but a boy had better learn his Latin over a man-trap than get the habit of smoking cigarettes." — *New York Medical Record.*

The use of cigarettes by young persons cannot be too severely condemned. They are made of the meanest materials, and often "doctored" with refuse substances, and even forms of opium, in order to give some bulk and "tone" to the originally cheap and filthy material. Cigarettes are so common and so cheap, that their use by thousands of young persons has become a serious matter.

Here is one bit of advice for you to remember all the days of your life : Do not smoke or chew tobacco if you wish to keep strong and well, and to succeed in life.

169. The Use of Tobacco from a Moral Point of View. — The effect of tobacco on the moral nature often shows itself in a selfish disregard for the rights of others. The smoker has no right to make the air about him unfit for others to breathe, with his tobacco smoke ; he has no right to puff his smoke into the faces of people on the streets, or to thus pollute the air of public places which others are obliged to share with him.

The fact that he does this knowing that to many people the smoke of tobacco is offensive, and that some are even made sick by it, shows his lack of refinement as well as moral sense. Other evidences of the same character are the filthy habits of spitting on sidewalks, floors, stoves, and other objects which characterize both smokers and chewers of tobacco and disgust all cleanly people.

It is no mark of friendship or courtesy to offer a person a cigar or cigarette, for it is virtually asking him to take what will be to him more or less of an injury instead of a benefit.

**170. Opium.** — Opium, one of the most powerful poisons, is the dried juice of the white poppy. It has the power of relieving pain, and producing a kind of sleep. Morphine is a white crystalline powder made from opium. Laudanum is made by macerating powdered opium in dilute alcohol. Paregoric is a weak tincture of opium combined with camphor and aromatics. Dover's powder is made of opium and ipecac.

The various forms of opium are very generally used in patent medicines. They form the "soothing" basis of liniments, "cough-killers," "soothing-sirups," stomach-bitters, cholera-mixtures, and countless other preparations, which people are eager to buy, hoping to get relief from some real or fancied disease.

Thousands of tons of opium are smoked and chewed, like tobacco, in distant parts of the world, especially in China, causing great degradation and misery among the people. This habitual use of opium is, however, a common vice everywhere. Over half a million pounds were imported into this country in one year recently, and a large proportion of this enormous quantity, it is claimed, was consumed by "opium-eaters."

Opium leaves its after-effects in dryness of the mouth, thirst, nausea, constipation, and a dull headache. In large doses, there are giddiness and stupor. A person becomes motionless and insensible to outward impressions ; he lies quite still, with the eyes shut, and the pupils contracted ; and the whole expression of the countenance is that of deep repose. As the poisoning advances, the features become ghastly, the pulse feeble, and the muscles greatly relaxed; and, unless help is procured, death speedily ensues. If the person recovers, the insensibility is succeeded by sleep for one or two days, followed by nausea, vomiting, and loathing of food.

Eaten or smoked habitually to satisfy a craving for it, opium makes a living death for its victim. A person may begin in the most innocent way to use a little opium to relieve pain: little by little, the meshes of this fascinating narcotic are woven about him. The opium relieves the suffering, but the worst of it is, a person cannot leave off its use without the greatest effort. A craving is stimulated which no one can realize unless he has once been within the destructive toils of this drug. It is untold misery to quit it, and sure death to keep on using it.

This habit of taking opium completely changes its victim. A man, once upright and honest, will lie, cheat, defraud, or even commit murder, to satisfy his awful craving for this baneful drug. Promises and resolutions to stop its use may be honestly made, but are no more binding than ropes of sand. He seems to be worked upon by some powerful charm.

The deepest melancholy settles on the opium-eater; and life, once full of joy and happiness, is indeed a heavy burden. He would gladly commit suicide, but he cannot summon the will-power to do the deed.

Meanwhile all the vital organs are slowly and surely losing the power to do their work. The fat disappears; the muscles grow weaker; the stomach and bowels fail to act; the skin becomes dry, yellow, and shrivelled, like yellow-leather parchment; and death at last puts an end to his misery.

**171. Practical Points about Opium.** — The so-called soothing-sirups, cough and cholera mixtures are often given to infants and young children. All of them contain more or less of some form of opium.[1] The child is simply drugged, and not cured, however "soothing" the effect may be. The only safe rule is, never to put opium on the list of home remedies.

A good physician often hesitates to use this drug with children, for he knows how dangerous and uncertain the effect may be. The practice of rubbing the gums of teething-children with paregoric, putting laudanum into a child's aching tooth or ear, giving either preparation for "summer complaints," and other household ways of using opium, is dangerous.

Nine times out of ten, some simple and safe remedy will answer every purpose. Never rub any form of opium powder upon any abraded surface to relieve pain. It is rapidly taken up by the blood. "Dover's powder" is not a home remedy: a single grain has killed an infant. Remember that laudanum that has been kept in the cupboard for a long time may become stronger, on account of the evaporation of the alcohol.

Remember that children as a rule, and many grown-up people too, are very susceptible to the action of opium. It is well known that a child may be drugged by the milk of a nurse who has taken opium. Young

[1] "It is very certain that many infants annually perish from this single cause."
— REESE'S *Manual of Toxicology.*

boys have been made stupid, getting even the "pin-
hole pupil,"[1] by smoking cigarettes that have been doctored with opium. Poultices saturated with laudanum
are by no means safe. They are especially dangerous
in the case of infants.

The only safe rule, therefore, is, to let everything
that has opium in it severely alone. Above all things,
do not give to others, especially children, any opium-
mixture that has been prescribed or used by some
one else. It is better to endure pain.

**172. Chloral.** — Chloral is a powerful drug, capable,
in small doses, of producing quiet sleep. This action
is probably due to its direct effect upon the brain. In
full doses it depresses the action of the nerve-centres
of the brain and spinal cord. In large doses, there is
paralysis of the nerve-centres, causing death. Because
chloral is known to induce sleep, especially in those
who suffer from excessive mental strain, or from
anxiety, or other like cause, it has come, of late years,
to be often used without a physician's advice.

Like all narcotics, the dose must be steadily increased
to get the required effect. The "chloral habit" is
soon formed, and the person becomes a slave to a dangerous drug. Without it, the chloral-eater cannot
sleep: with it, his digestion is sadly out of order. He
suffers from dyspepsia, shortness of breath, and palpitation of the heart. The habit begets carelessness in
its use; and the fatal dose is so uncertain, that chloral-
eaters often die from an overdose.

---

[1] One of the physiological effects of opium is to contract the pupil of the
eye. Hence the name of "pin-hole" is given to the pupil contracted by a large
dose of any opiate.

The only safe rule is, never to touch so powerful, uncertain, and dangerous a drug. It is better to consult the family physician.

**173. Other Powerful Drugs and Narcotics.** — Regarding chloroform, ether, hasheesh or Indian hemp, and other narcotics, it may be said, once and for all, that they should never be used, even in the smallest doses, unless under medical advice. They are dangerous agents at all times, and are used with great caution, even by physicians. Persons who get into the habit of tampering with such powerful drugs run the ever-present risk of killing themselves by an overdose.

Inhaling ether or sniffing chloroform on a handkerchief, as a household remedy, to relieve spasms of pain, should be always severely condemned. Familiarity with these and all other powerful drugs and narcotics is sure to beget carelessness on the part of those who use them on themselves.

Since that far-reaching epidemic known as influenza, or "la grippe," made such sad havoc in the past few years with the lives and health of people all over the world, a new class of uncertain but powerful drugs has come into extensive use, even in the household.

They are known by various fanciful and scientific names. Let them severely alone. They are too powerful and uncertain drugs for household use. Remember that your family physician will prescribe them only with the greatest caution.

# CHAPTER XII

## THE SPECIAL SENSES

**174. Sensation.** — In the preceding chapter we have seen that all nerves belong to two classes : —

1. **Motor nerves,** which control the action of the muscles of the body.

2. **Sensory nerves,** which carry a variety of impressions from all parts of the body to the brain.

When the brain receives a certain impression by means of some sensory nerve, we become conscious of a feeling, or **sensation.**

Exactly how we become conscious of the thousand and one sensations felt by all, is one of the many mysteries connected with the human body. Some sensations, such as those of faintness or fatigue, spring up within us in some mysterious way, without any apparent cause. Most sensations, however, are produced by some outward agency.

Thus, if we hear a child cry, or a bird sing, we have a sensation of sound. If we are pinched, tickled, or struck, we have a sensation corresponding to these acts. If we put a piece of sugar on the tongue, hold a rose at the nose, or prick the skin with a pin, certain sense-organs receive the impressions, and the sensory nerves carry them to the brain, where we become conscious of a sensation.

Sensations may be those of pleasure or pain. Indeed, the same agent may be either pleasant or painful, according to its degree of intensity. It is pleasant to hold out the hands before a glowing fire after a cold ride. Hold the hands too closely, and the sensation of pleasure is changed in a moment to one of pain.

All sensations are produced by three distinct organisms : —

1. An organ specially adapted to receive the stimulus from the outer physical agent.

2. An incoming, or sensory, nerve to carry the impression from the sense-organ to the brain.

3. The brain itself, some part of which converts the impression into an actual sensation to the mind.

Take, for instance, the sense of hearing. The waves of sound pass through the air from the outer physical agent. The ear is the organ specially adapted to receive impressions from it. The nerve of hearing, or auditory nerve, forms the connection between this organ and the brain, where the consciousness of sensation actually takes place.

**175. The Five Special Senses.** — There are five ways in which the brain becomes acquainted with what takes place in the outer world. In other words, there are certain sets of nerves which carry special sensations to the brain.

We are said to have five special senses,[1] or "gateways of knowledge," — **touch, taste, smell, hearing,** and **sight.**

---

[1] There is another sense, commonly called the *muscular sense*, which enables us to judge the weight of different bodies, according to the muscular effort required to lift or hold them. This sense becomes so highly developed with use, that shopkeepers, and others who sell various articles by weight, will often tell you the weight of a body by simply balancing it in their hands.

**176. Touch, or Feeling.** — The sense of touch is the most widely extended of all the senses, and perhaps the simplest. It has its seat in the skin all over the body, and in the walls of the mouth and nasal passages. By this sense of touch we tell whether a body is hard or soft, hot or cold, rough or smooth. We have been told in chapter x. about the thousands of tiny hillocks called *papillæ*, which form rows of ridges very thick on the tips of the fingers.

Now, besides a tiny artery and vein, finer than the finest hair, there is in each papilla the end of a sensory nerve. Where the sense of touch is most delicate, the papilla is found to contain a little oval bulb called the touch-corpuscle. All parts of the body do not have this sense of touch in an equal degree.[1] It is most delicate in the tip of the tongue, the tip of the fingers, and the edge of the lips, and least delicate in the middle of the back.

**177. Taste.** — The tongue is the principal organ of taste. It has two coverings, — a deep, sensitive layer, and an outer layer. When the stomach is out of order, and in certain diseases, this outer layer is covered with a whitish or yellowish matter. We generally speak of the tongue at such a time as being coated.

The deep layer is raised up, like the true skin, into

---

[1] There is no sense so capable of improvement as that of touch. The female silk-weavers of Bengal are said to be able to distinguish, by the touch alone, twenty different degrees of fineness in the unwound cocoons, which are sorted accordingly.

The sense of touch may be said to belong to every animated being, and is one great characteristic of animal life. In many animals the tongue is the instrument of touch as well as of taste. Certain animals, in addition to the tongue, have special organs of touch, as the whiskers of the cat and rabbit.

tiny hillocks, or papillæ, which are abundantly supplied with delicate nerve-fibres from two great nerve-branches leading from the brain. These are the nerves of taste.

The tip and the back of the tongue are supplied with different nerves. The papillæ are small, and pointed on the tip, and arranged in the form of a letter V, with the point towards the back.

Hence it makes a difference in the taste, whether we put a substance to be tasted on the tip or back of the tongue. Thus, alum tastes sour on the tip, and has a sweetish taste on the back part, of the tongue.

In certain animals, these last-mentioned papillæ are very largely developed, and

FIG. 68. — Upper Surface of the Tongue, showing the Papillæ.

give a roughness to the tongue. We have all, no doubt, noticed how rough a dog's tongue is. It is this which enables the dog to strip off the flesh from a bone by simply licking it; while the lion by the same means strips the skin from his victim with one stroke of his tongue.

In this sense, as in the other special senses, the nerves receive the impression, and carry it to the brain, where it is perceived, and gives rise to the sensation of taste.

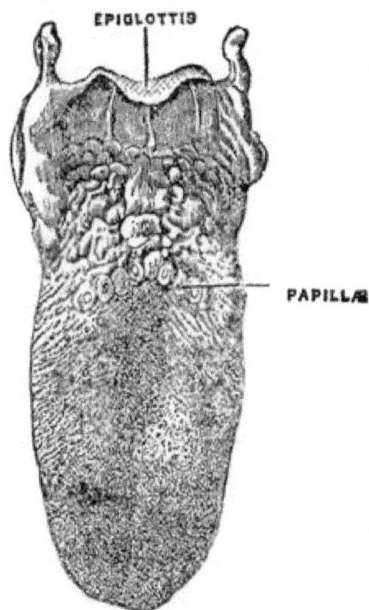

**178. Smell.** — The seat of the sense of smell resides in the cavities of the nose, into which the nostrils open, and which open behind into the pharynx, or the back part of the mouth. The walls of the nasal cavities are lined with a thick, velvety membrane, over which the nerves of smell are distributed. This membrane is kept continually moist by a fluid which it secretes. At the beginning of a cold in the head, this membrane becomes dry and swollen, and the sense of smell is greatly lessened.

It is in the roof of the nasal cavities that the sense of smell is most acute.[1] Hence, when we wish to detect a faint odor, we sniff up the air sharply. By doing so, we cause a rush of air into the higher parts of the cavities, where some of the floating particles come into contact with the nerves of smell.

The sense of smell is Nature's sentinel to guard us against taking improper food into our stomach, and impure air into our lungs. By this same sense, we are able to detect delicate and fragrant odors, which add much to the comfort of our daily living.

**179. The Sense of Hearing.** — We come now to a special sense, which does not tell us what is going on in the outer world by actual contact, as in touch or

---

[1] The sense of smell varies very much in different individuals. Among civilized people, it is rather defective, while in savage races it is most acute. We are told that the South-American Indians can detect the approach of a stranger, even in a dark night, by their sense of smell, and can also tell whether he is white or black. Many animals are more highly endowed with this sense than man. Thus, a dog will smell the footsteps of his master amid those of a hundred other people, and can trace him for miles, although he has been for hours out of sight. Pointers also scent game at a great distance.

taste, nor by particles of matter falling upon the end of nerves, as in the sense of smell. In this sense, impressions are produced upon nerves by wave-like vibrations in the surrounding air.

This is the sense of hearing, only second in importance to the sense of sight. All sounds are produced by the vibration of some body in the atmosphere. The

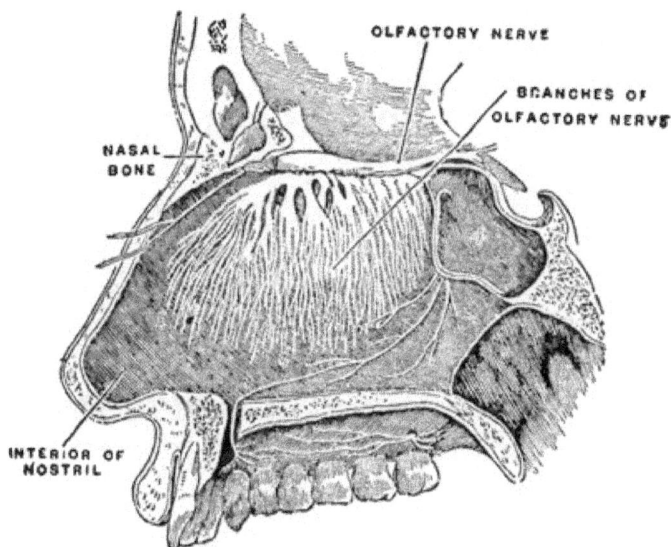

FIG. 69. — Sectional View of the Nose.

body sends out these vibrations to the surrounding air, which carries forward a series of waves in all directions.[1]

These air-waves are received, and the impression

[1] We shall understand these air-waves better if we throw a stone into a pool of water, and watch the result. We see a series of tiny circular ripples gradually spread themselves over the surface of the water from the spot where the stone fell. This exactly represents the waves of sound caused by the vibration of bodies in the air. These air-waves travel with wonderful rapidity. The usual velocity of sound is about eleven hundred feet a second.

made by them is sent to the brain, by a special apparatus, called the organ of hearing.

The organ of hearing is lodged in the thick inner bones forming the base of the skull. The ear, the organ of hearing, is far more complicated than any of the organs of sense yet described.

It is divided into three parts, — the *outer*, the *middle*, and the *inner* ear.

**180. The Outer Ear.** — The **outer ear** consists of a broad plate of gristle, shaped somewhat like a shell, commonly called "the ear;" and of a tube about an inch long, called the **auditory canal.**

This canal is in the solid portion of the temporal bone. It is protected by fine hairs, and by a set of glands which secrete ear-wax, a bitter substance, which serves to moisten the parts, catch particles of dust, and keep away small insects.

The inner end of this canal is closed by a membrane stretched tightly across it. It resembles the parchment stretched across the end of a drum, and is known as the membrane of the drum of the ear. It is thin and elastic, but may be easily broken by a blow, or by pushing anything into the ear. If once broken, this delicate membrane cannot be repaired, and deafness results.

**181. The Middle Ear.** — The **middle ear** is a small cavity hollowed out in the temporal bone, between the membrane of the drum and the inner ear. This cavity is the "drum" of the ear.

The most curious feature of the drum is a string of three of the tiniest bones, which stretch across it. They are sometimes called the hammer,[1] anvil,[2] and

[1] *Malleus.*　　　　　[2] *Incus.*

stirrup,[1] bones from some fancied resemblance to these articles.

The hammer-bone is fastened by its long handle to the membrane of the drum. The round head of the hammer-bone fits into the anvil-bone. Next to the

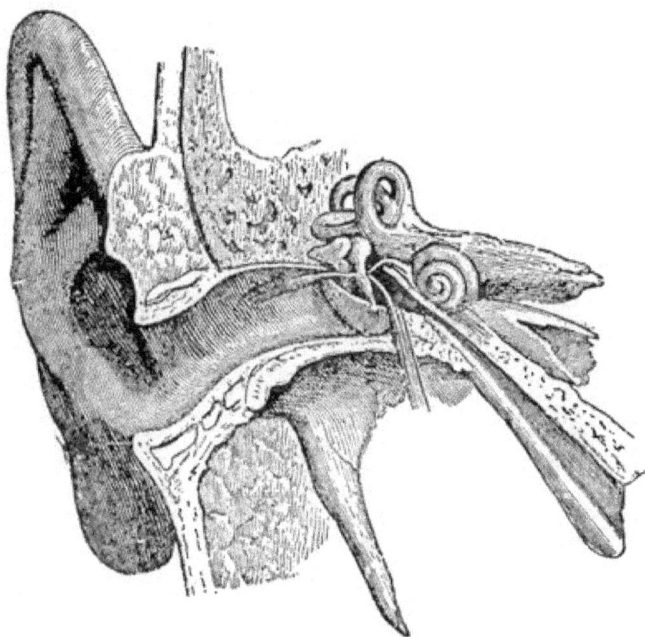

Fig. 70. — Section showing the different parts of the Ear.

anvil is the stirrup-bone, which fits into a little oval window in the opposite wall of the chamber, or drum.

In the floor of this chamber is the opening of a passage called the **eustachian tube**.[2] This tube is about an inch and a half long, and leads into the throat. It

---

[1] *Stapes* (the smallest bone in the human body, being one-eighth of an inch long).

[2] Named after Eustachi, a famous Italian anatomist, who died more than three hundred years ago, and who first described it.

allows air from the throat to enter the drum, and serves to keep the air on both sides of it at a constant and even pressure.[1]  To hear perfectly well, this passage must be kept open for the air to pass in and out.

During a severe cold in the head, or a sore throat, the lining of the tube may be inflamed and swollen. This gives a stuffed feeling and an alteration of sounds in the ears.  This should not be neglected if it does not pass off in a few days.  Should it continue, it will show that more than ordinary swelling has occurred in the tube; and it will be found that the hearing is slightly impaired.  Usually, as the cold passes off, this peculiar feeling in the ears passes away.

**182. The Inner Ear.** — The inner ear is one of the most delicate and complex pieces of mechanism in the whole body.  It is that part of the organ of hearing which perceives the impression of sound, and carries that impression directly to the brain, where it gives rise to the sensation of hearing.  It is really a bony case filled with a watery fluid, in which float the delicate parts of the inner ear.

The inner ear consists of three distinct portions, — the hall-way,[2] the canals,[3] and the snail's shell.[4]  It is enough to say that these are winding channels and spiral tubes hollowed out in the solid bone.  The whole

---

[1] This passage is ordinarily closed ; but when the orifice in the throat is opened, as in the act of gaping or swallowing, the air rushes into the cavity of the tympanum.  Close the mouth, and pinch the nose, and then try to force air through the latter.  Air is thus forced through the eustachian tube.  A distinct crackle or clicking sound is noticed, due to the vibration of the membranes, and the movements of the little bones of the ear.

[2] Vestibule.          [3] Semi-circular canals.          [4] Cochlea.

system of passages is known as the "labyrinth." It is important to remember, that there is a continuous connection between all the passages of the inner ear, and that all the winding tubes and chambers enclose and

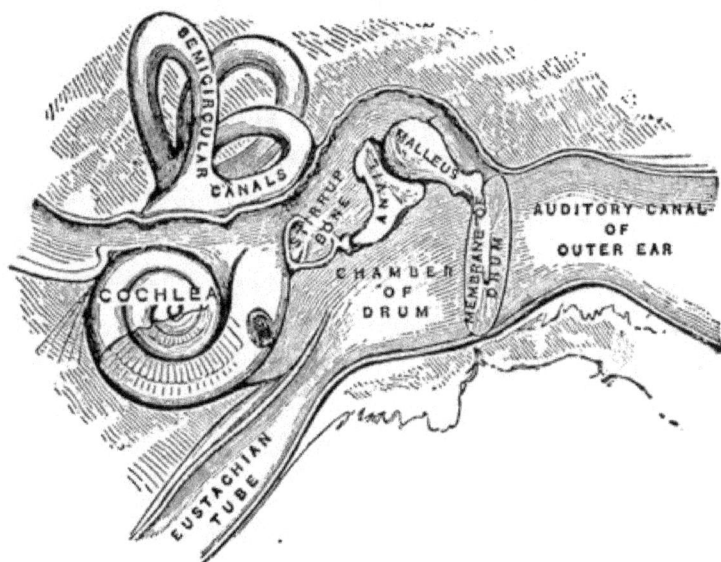

FIG. 71. — The internal mechanism of the Ear — greatly enlarged.

protect a delicate bag of membrane of exactly the same shape as themselves.

The **auditory nerve,** or nerve of hearing, passes from the brain, through a little hole in the solid bone of the skull, to the inner ear, and is spread out on the lining membrane.

**183. How we Hear.**— Let us try and understand a few of the simplest principles of this wonderful mechanism of the ear. A bell is rung, or a gun fired. The vibration is communicated to the atmosphere around it, and passes away in air-waves from the sounding body, as the waves ripple the surface of a pond after a stone has been thrown in.

The air-waves go into the outer ear, and strike upon the stretched membrane of the drum, and cause it to vibrate. At every vibration of the membrane, the head of the hammer-bone strikes upon the anvil-bone, drives it forward, and pushes the foot-plate of the stirrup-bone in and out of the oval window on the inner wall of the chamber.

The watery fluid in the inner ear washes tiny grains of sand against the membranous bag, and so strikes the ends of the auditory nerves. These nerves flash the impression they have received to the brain. It is in the brain itself that the sensation of hearing takes place.

**184. Care of the Ear.** — The ear is a very delicate organ. It is often carelessly and ignorantly tampered with, when it should be let alone. It is often neglected when skilled treatment is urgently needed.

The ear-canal should never be rudely or hastily washed out, either with a syringe or a wash-rag. The utmost gentleness in washing out the ear is all that is necessary for cleanliness. Children's ears should never be pulled or boxed.[1] Even a slight blow has resulted in serious trouble.

Never use ear-picks, ear-spoons, the end of pencils or

---

[1] Thus, a "box on the ear," a blow on the ear or head of any kind, or a fall may be the direct cause of permanent deafness. Sometimes the deafness is total as well as sudden.

Deafness from concussion, as a tumble down-stairs in childhood, or from a severe blow of any kind on the head, may be the foundation of deaf-dumbness, if the injury occur before the child has learned to talk.

pen-holders, pins, hair-pins, tooth-picks, towel-corners, etc., to pick, scratch, or swab out the ear-canal. It is a foolish, needless, and dangerous practice. There is always risk that the elbow may be jogged in many ways, and the pointed instrument pushed through the drum-head.

Let the ear-wax take care of itself. The skin of the ear grows outward, and the extra wax and dust will be naturally carried out if let alone. Never drop sweet-oil, glycerine, and other fluids into the ear, with the idea that the ear is cleaner for them. They do no good, and often irritate the ear. Never advise or allow any of the many nonsensical things so commonly used, to be put into the ears to cure deafness. Cotton wads may be gently put into the ears to shield them from the cold, or may be worn in swimming or diving, to keep the water out. Diving into deep water, or bathing in the breakers, often injures the ears.

One should never shout suddenly in a person's ear. The ear is not prepared for the shock, and deafness has occasionally resulted. If the eustachian tube is closed for the time, a sudden explosion, noise of a gun or cannon, may burst the drum-head. Soldiers during heavy cannonading open the mouth to allow of an equal tension of air.

Flies, bugs, ants, and the like, sometimes crawl into the ear. This may cause some pain and fright, and perhaps lead to vomiting, and even convulsions with children. A lighted lamp put at the entrance of the ear will often coax insects to crawl out towards the light. The ear may be syringed out with a little

warm water.   Drop in a few drops of molasses or warm
sweet-oil.

When the ears run for any cause, it is not best to
plug them with cotton wads.   It only keeps in what
should come out.   Very cold water should never be used
in the ears or nostrils if it can be helped.   Use only
tepid water.   Do not go to sleep with the head on the
window-sill, or in any position that may expose the ears
to a draught of cold or damp air.

**185.   The Sense of Sight.** — Sight, or vision, is a
wonderful thing when we come to think about it.   That
we have a means by which we can tell what is going on
in the outside world, is a precious gift.   We watch a
huge balloon from the time it leaves the ground till it
is a mere black speck in the atmosphere above us.   We
follow the vessel sailing along on the dim horizon, and
the next instant we are reading the fine print of a newspaper.   We recognize the form, size, color, and distance
of thousands of different objects in nature every day of
our lives.

This sense is so woven into the countless acts of our
every-day affairs, that we scarcely appreciate this marvellous gift, so essential, not only to the simplest matters of comfort, but also of prime importance in the
culture of the mind and the higher forms of pleasure.

Sight is well held to be the highest and most perfect
of all the senses.

**186.   The Eye.** — The eye is the outer instrument of
sight, and is a most beautiful and ingeniously built
machine.   This little organ, only about an inch in

diameter, is in reality one of the greatest wonders in nature.

The eyeball is lodged in a hollow cone made up of seven little bones, with the base pointing outwards and forwards. This eye-socket, or orbit, is well protected on its edges by the dense and strong bones of the head.

The eyeball rests in a soft and elastic bed of fat, which supports and protects it, and at the same time allows it to move in all directions as freely as if it floated in a dense fluid.

**187. The Coats of the Eye.** — The walls of the eyeball are made up of three distinct coats, or coverings.[1]

The outer covering is the tough coat (sclerotic), which is generally spoken of as the "white of the eye." It is one of the toughest and strongest membranes in the body, and is intended to protect the delicate structures within. It is so strong, that a sheep's eyeball may be trodden on with all the weight of the body, without being made to burst by the pressure.

This coat gives place in the front of the eye to a transparent circular plate, just as in a watch the gold or silver case gives place to a glass crystal over the face. This transparent plate is the cornea, meaning "a horn."

It forms a kind of rounded window for the eye.[2] The polish and transparency of the cornea are kept up by frequent, unconscious winking, which keeps it moist and free from dust.

The second coat (choroid) is much more delicate in

[1] 1, *Sclerotic;* 2, *Choroid;* 3, *Retina.*

[2] The cornea can be best seen by looking at it from the side, or by seeing the reflection of a window-sash upon its surface. We usually look directly through it, and see only the colored iris behind.

structure, and is rich in blood-vessels and nerves. It is lined with a thick black coating, designed to absorb the surplus rays of light, which would otherwise cause a blurred or confused vision.

The retina, meaning "a net," the innermost coat of the eyeball, is an extremely delicate film, which covers the inner surface of the choroid. It is a sensitive network, made up of fibres proceeding from the optic nerve, spread out over the inner surface of the eye.[1] It is the retina that makes the eye sensitive to light, so that we may well call it the actual or essential organ of sight.

**188. The Inside of the Eye.** — To get a clear idea of the inside parts of the eye, let us imagine an eyeball cut through the middle from above downwards. Let us now start in front, and go backwards. We shall first see the cornea, which has just been described.

We now reach a space called the front chamber of the eye. In this chamber, and behind the cornea, hangs down a curtain, the **iris,** meaning "rainbow." It has a hole through its centre called the **pupil,** which appears as if it were a black spot. It is this curtain which gives the color to the eye. The iris has muscular fibres which expand and contract, and thus make the pupil larger or smaller, according as the light is bright or dull.

When the light is very strong and brilliant, the iris spreads its curtain farther over the pupil in order to

[1] The retina is a film only 1-120 of an inch thick, yet ten distinct layers have been described.

shut out some of the rays. When the light is faint, the curtain is drawn back from the pupil in order to admit as many rays of light as possible. The black

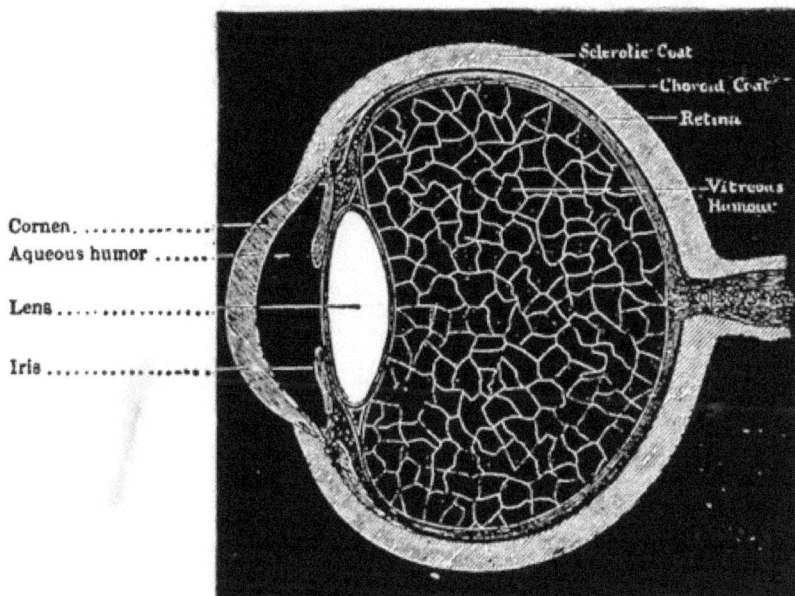

Cornea

Aqueous humor

Lens

Iris

Sclerotic Coat

Choroid Coat

Retina

Vitreous Humour

FIG. 72. — Section of Eyeball.

appearance of the pupil is due to the thick black coating which lines the inside of the choroid. It is like looking through a small window into a room in which the walls are hung with black paper.

Just behind the iris is a clear, transparent, jelly-like body, called the **crystalline lens.** It is convex, or rounded, both back and front, and is about one-third of an inch in diameter. It is shut up in a kind of transparent bag, and is held in its place by a number of little bands. The crystalline lens separates the front chamber of the eye from the back chamber.

The front chamber is filled with a clear, watery fluid called the **aqueous,** or watery, humor.

The back chamber also contains a jelly-like fluid called the **vitreous,** or glassy, humor.[1] Both the aqueous and the vitreous humors exercise some nutritive action on the adjacent parts, and help to support and keep tense the coats of the eyeball.

**189. Mechanism of Vision.** — Let us now trace the course of the rays of light going from any luminous body — a lighted candle, for example — through the different parts of the eye. Imagine the candle to be placed at about the ordinary distance of distinct vision (about ten inches) in the front of the eye.

Some of the rays fall on the outer coat, or the " white, of the eye," and, being reflected, take no part in vision. The more central rays fall upon the cornea. Some of these are reflected, giving to the surface of the eye its beautiful, glistening appearance.

Other rays pass through it, are brought nearer by refraction, and enter the aqueous humor, which also slightly bends them. Those which fall on, and pass through, the outer part of the cornea, are stopped by the iris, and are either reflected or absorbed by it. Those which fall upon its more central part pass through the pupil.

Now, if the rays of light passed directly to the retina, they would pass in parallel lines, and produce

[1] This glassy humor, even in healthy eyes, contains little bodies called "flitting flies," which sometimes frighten people who notice them, greatly magnified, moving up and down like strings of tiny bright beads. They are more plainly seen when looking at the bright sky or a white ceiling. They are in every eye, and can do no harm.

the impression of light, but everything would be dim and confused. Therefore it is necessary that the rays coming from any object should be brought together (converged) by being bent (refracted). That is, they must be refracted, and brought to a focus.

This is done to a certain extent by the cornea and the fluids, or "humors," of the eye, but mainly by the crystalline lens. This is a familiar fact in the use of any optical instrument. The boy changes the focus of his spy-glass by pulling its tubes in or out. When the lady looks from a distant to a near object with her opera-glass, she changes the focus by turning the adjusting-screw.

Now, it is the duty of the crystalline lens to bring the rays of light together as they pass through it, and to bring them to a focus on the retina. A tiny muscle, called the ciliary, or hair-like muscle, does for the eye what the adjusting-screw does for the opera-glass. As it contracts and relaxes, the elastic lens, held in its place by delicate cords, becomes more rounded or flatter, just as we may wish to look at distant or near objects.

Suppose we wish to look at an object close at hand. The little muscle contracts; and the layers of membranes, between which the lens lies, are the least bit relaxed; and, because it is elastic, the lens becomes more rounded, or convex, thus bringing the rays of light to a proper focus. If we wish to look at any object a long distance off, the membranes are drawn tight by the muscle to which they are tied, and the lens is flattened.

In other words, our eyes adjust, or *accommodate*, themselves to the varying distances of objects ; just as the photographer pushes in or out the lens of his camera with his hand, so as to bring it farther or nearer from the surface which receives the image.

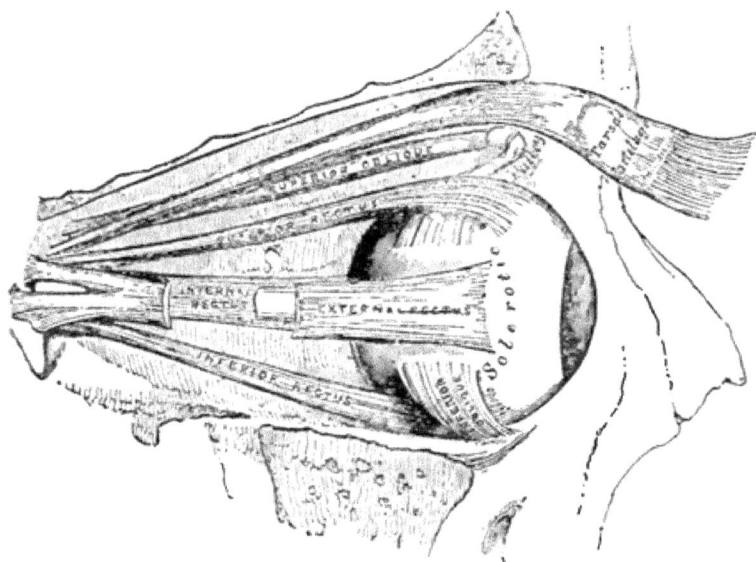

FIG. 73. — Muscles of the Eyeball.

The rays of light are now ready to be brought to a focus on the sensitive retina. The iris has regulated with wonderful skill the proper amount of light ; and the lens, with the greatest exactness, has focussed the rays on the retina. The dark surface of the middle coat acts to absorb the excess of light, and thus prevent the reflection which would disturb accurate vision.

As a result, an exact but inverted image is formed on the retina. The impression is carried to the brain by the fibres of the **optic nerve,** which are spread out on this wonderfully sensitive membrane.

**190. The Muscles of the Eye.** — The eyeball is rolled and moved about by six muscles. They spring from the back part of the bony orbits, and are fastened to the front part of the eyeball by means of tendons. Four of these muscles — the recti or straight muscles — move the eye up or down, and to the right or left. The other two, the oblique, are so fastened that they turn the eyeball in one direction or another. Sometimes one or more of these muscles pull unequally, and a "squint" or "cross-eye" is produced.

**191. The Eyelids and Eyebrows.** — The eye, delicate organ that it is, is protected and kept clean by its **eyelids, eyelashes,** and **eyebrows.**

The eyelids are thin, flexible covers, or shutters, which protect the front of the eyeballs. They are thin plates of gristle covered with skin. They are lined on the inner side with a very delicate membrane called the **conjunctiva,** because it is also joined to the eyeball by a fold.[1] It pours out an oily fluid to prevent friction between the surfaces.

This fluid, together with a constant flow of tears, keeps the cornea moist, and free from dust. The edges

---

[1] This membrane covers the front part of the sclerotic, the whole of the visible portion, and, lining the walls of the tear-duct, becomes continuous with the mucous membrane of the nose and throat, and, therefore, usually takes a part in a "cold in the head."

It is, ordinarily, nearly white, with perhaps a few of the larger blood-vessels seen winding through it, but is very easily congested by local injury, or inflammation, particularly of the head, as in the case of the "blood-shot" eye. The characteristic yellow tinge of the eye in jaundice is due to the coloring-matter of the bile deposited in the conjunctiva.

of the eyelids are provided with a fringe of fine hairs, the eyelashes, which help shade the eye, and shield it from dust.

The eyebrows form a protecting and shading ridge over the eyes ; while the thick fringe of hairs, arranged somewhat like the straw on a thatched roof, prevents the sweat from rolling into the eyes as it trickles down the forehead.

**192. The Tears.** — Nature provides a special fluid to protect the eye. This fluid is called the tears.

The tear-apparatus consists of the gland for secreting tears, and the passages for draining them off. This gland pours out the tears over the surface of the eye, moistening it by the winking of the lids, and making its movements easy.

The tears are now taken up, and carried off, by two fine tubes, one in the upper and one in the lower lid, which unite, and form the tear-duct, which leads into the nose. The ordinary flow of tears is thus drained off.

Local irritation or mental emotion may excite an excessive flow, which the canals cannot carry off. The tears then overflow, and run down the cheeks. This is called crying, or weeping. Nature kindly greases the edges of the eyelids, to prevent, to a certain extent, the overflow of tears.

**193. Color-blindness.** — Color-blindness is the inability to tell certain colors. It is sometimes produced by sickness ; but generally it exists at birth, and is often hereditary.[1] This defect of sight is quite common.

---

[1] Some years ago a physician gave a history of color-blindness which had existed in five generations of his own family.

Out of the many thousands that have been examined, it is found that four or five per cent are color-blind. A person may be color-blind, and not know it until the defect is accidentally revealed.

This is a matter of the utmost practical importance to those employed on railroads, vessels, and other places where colored signals are used. Some are only partially color-blind, while others are wholly so. The most common form of color-blindness is that in which one fails to distinguish red.

**194. Near-sight and Far-sight.** — All eyes are not perfect. **Near-sight** is a common defect of vision. In the healthy eye, the rays of light are brought to a focus on the retina. But in some eyes the image is blurred; the outer coat, bulging backward, making the eyeball a little too long, thus bringing the rays of light to a focus before they reach the retina. A person with such an eye is said to be " near-sighted."

It may exist at birth, and is very often hereditary. It is acquired by overstraining the eyes, in reading too fine print, by reading by a dim or imperfect light, and in many other ways. This defect is common with those who use their eyes much in reading, writing, and study. Sailors, farmers, and others who work out-doors, are rarely near-sighted. There has been found to be a steady increase of near-sightedness during the period of childhood and youth, especially among school-children. The statistics are so startling, that the subject is really one for the most serious consideration. Near-sightedness, in many cases, is a serious disease, which demands the most careful and skilful treatment by some good oculist. Teachers can do much to check the ill effects of this disease among their pupils.

In **far-sight** the eyeball is too short, or the lens may be too flat. The rays of light are not brought to a focus. As a result, the image is blurred. The oculist can easily advise the proper kind of glasses to correct both of these defects of vision.

**195. Care of the Eyes.** — The eye is an exceedingly delicate and sensitive little machine. It is an easy matter to get it out of order, and very tedious to restore it to health again. The eyes are often weak after certain sicknesses. The utmost care must be taken of them during and after an attack of measles and scarlet fever.

The habit of reading the daily papers, with their blurred and indistinct type, in a car or carriage, is a severe strain on the eyes. We have to change the focus to suit the variation in the distance between the eyes and print caused by the constant jarring. It is a dangerous practice to read in bed at night, or while lying on a sofa or lounge in a darkened room. The outer muscles of the eyeball are put to a severe strain. The small type, poor paper, and press-work of the many cheap editions of popular books, now so commonly read, are very frequent causes of weak and diseased eyes.

The direction in which the light comes, is an important matter. The worst direction of all is that from in front. The direct light should fall upon the print from above, and from the left side. The nearer an artificial light can come to a mellow daylight, the better. The flickering light from the ordinary "fish-tail" gas-burner is always bad. Its light should be regulated by shades and globes. Recently invented round and hollow burners, used with kerosene oil, give an excellent light.

After reading steadily for some time, we should rest the eyes by looking at some distant object, even if only for a minute. A person should never read, write, sew, stitch, or otherwise use the eyes, when they tingle or smart, or the sight is dim or blurred. The eyes are weary, and need a rest.

Using the eyes at dusk, or by artificial light in the early morning, is apt to lead to serious disorders of vision. The eyes should never be rubbed, or the fingers roughly poked into them at any time, and much less when they are irritated by getting some foreign substance into them. The sooner it is removed, the better. Rubbing the eye, or pulling the eyelids, only makes a bad matter worse. When the eyes smart after going to bed, or tingle, and are "bloodshot" on getting up the next morning, it is safe to conclude that they have been overtasked, and need rest.

It is not a wise economy to tamper with one's eyes when they are ailing. Better to do nothing than do the wrong thing. If a few days of rest do not give any relief, a good oculist should at once be consulted.

**196. Effect of Alcohol and Tobacco upon the Special Senses.** — We have already learned that alcohol in a general way dulls or weakens the nerves. Hence it must have a similar effect upon the nerves of sensation. As a matter of fact, alcoholic liquors and tobacco dull the senses, and even provoke changes in the sense-organs themselves.

Thus, strong ·drink inflames the throat, and then the Eustachian tube, and so indirectly injures the sense of hearing. It lessens the quickness and the acuteness of hearing. Tobacco, also, often irritates the lining of the throat and nose. This inflammation is apt to extend up the tube and involve the delicate structure of the middle ear, thereby causing some defect in hearing, such as a difficulty in hearing sounds that are very soft or very loud.

Alcoholic liquors often produce congestion of the eyes and an irritation of the delicate lining of the eyelids. Even in very small amounts alcohol lessens the acuteness as well as the quickness of vision. An immoderate use of strong drink occasionally produces disease of the retina and the optic nerve.

Tobacco smoke irritates the eyes. It may cause sharp pain and a redness of the eyeballs. Sometimes images are retained for some time after the eye has ceased to look at objects. This may be due to the effect of the poison of tobacco upon the retina. Smokers occasionally suffer from a feeble and confused vision, due to a partial paralysis of the optic nerve. It is also claimed, by careful students of the subject, that the weak and imperfect vision of many young people of our day is due to the smoking habits of their fathers.

Cigarettes are especially hurtful to the throat and the eyes. They produce an inflammation of the lining membrane. An atmosphere laden with tobacco smoke often irritates the lining of the bronchial passages.

# CHAPTER XIII

## EXCRETION

**197. Getting Rid of Waste Matters.** — Our bodies are never the same for a single moment. With every breath, and with every beat of the heart, they are ever changing. Wear and waste vie with growth and repair.

We eat food to supply the bodily engine with fuel, and breathe in oxygen to feed the furnace-fire. With a steady burning, but without light, the engine produces not only motion and heat, but uses a part of its own energy to make its own repairs. Not this alone, but it even gets rid of its own soot, clinkers, and ashes, which would otherwise clog the machinery, and finally stop it.

We have already learned that it is the ceaseless blood-current which carries fresh fuel, in the shape of the nutrient part of food, to the tissues, and is a kind of a sewer-stream that rids them of waste matters. The blood is ever being made rich by some things, and is ever getting rid of other things. The blood carries fuel to the tissues of the body. A slow burning takes place : the waste or ashes must be got rid of.

**198. Principal Waste Matters of the Body.** — What are these waste matters ? If we take a piece of beef, dry it, and burn it, we shall find that it is changed into four things, — water, carbonic acid, ammonia, and ashes. Now, this slow burning of our tissues is really the same thing. Hence, in whatever way the body is burned, or oxidized, whether in a furnace, or while it is' living, the end is always the same : —

**Water, Carbonic Acid,** a kind of ammonia called urea, and a small quantity of salts, or ashes.[1]

Besides the *water* which comes from the burning of the hydrogen of our food, we are drinking a great deal. We need to keep the tissues continually moist, to help dissolve the food, and also to wash the body inside, and cleanse it of its useless matters and impurities. As we wash the surface of the body to keep it clean, so Nature is ever giving our tissues a bath, to wash away their impurities.

The red blood-disks, as we know, are the tiny boats which carry the oxygen breathed in by the lungs along the blood-stream to every tissue. The tissues contain carbon; and in some mysterious way, the oxygen unites with the carbon, and *carbonic acid* is formed.

The tissues, like the muscles and nerves, yield nitrogen. A partial decomposition takes place; and the nitrogen is filtered out of the body, through the kidneys, in the form of *urea*. It is a peculiar substance, something like ammonia, only more complex. Neither carbonic acid nor urea is fit to build tissue, or to set free energy.

---

[1] The body is made up of nitrogen, carbon, hydrogen, and oxygen, with sulphur, phosphorus, and some other elements. The nitrogen and hydrogen go to form ammonia ; the hydrogen, with the oxygen of combustion, forms water ; the carbon, carbonic acid ; the phosphorus, sulphur, and other elements, go to form the various salts of the body.

**199. The Organs of Excretion.** — The process by which the body gets rid of its waste material is called **excretion,** meaning to separate from, or to sift out.

The chief organs of excretion are the **skin,** the **lungs,** and the **kidneys.**[1]

The functions of all three of these organs are closely allied. The organs differ very much in appearance, but are built on the same principle. The blood, as it passes through the numberless capillary channels in them, is purified by a sifting process. The products of excretion are, as it were, sifted from the blood, and finally removed from the body.

The **lungs** have also been described in a previous chapter. The duty of the lungs, as we have learned, is to excrete carbonic acid, watery vapor, and a small portion of worn-out animal matter.

The structure of the **skin** has already been described. The main function of the skin, as we have seen, is to rid the body of water and other matters making up the sweat.

**200. The Kidneys.** — These two important organs of excretion are of a brownish-red color, about four inches long, and two inches wide. They lie in the region of the loins, in front of the backbone, behind the intestines, one on each side, and are embedded in fat. A sheep's kidney is a familiar sight in a market. Now,

---

[1] "The three great channels, then, by which the blood purifies itself, by which it gets rid of its waste, are the lungs, the kidneys, and the skin. Through the lungs, carbonic acid and water escape; through the kidneys, water, ammonia in the shape of urea, and various salts; through the skin, water and a few salts. As the blood passes through lung, kidney, and skin, it throws off little by little the impurities which clog it, one at one place, another at another, and returns from each purer and fresher." — FOSTER'S *Physiology.*

the kidneys of man have the same shape, and look
like those of a sheep, only a little longer.

The name "kidney" is popularly given to a kind of
bean," the kidney
bean," because of
its resemblance to
the shape of a kid-
ney.

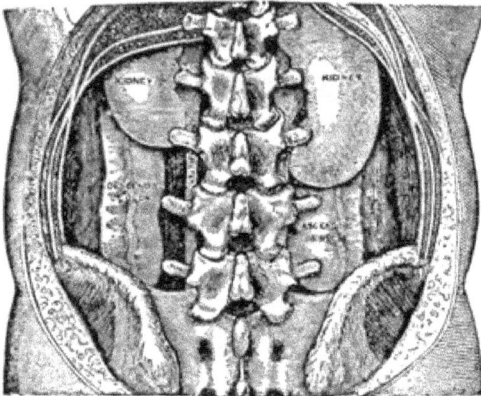

A kidney is a bun-
dle of long tubes,
not so very unlike
sweat - glands, all
bound together into
the kidney-shaped
mass, whose appear-
ance is familiar to
FIG. 74. — The Kidneys.

us.   The blood filters certain waste matters which become urine into these tubes, just as it secretes sweat
into the sweat-glands.

The urine is conveyed from the kidneys by two tubes,
called the ureters, to the bladder, from which it is cast
out of the body.   Thus the kidneys purify the blood by
carrying off urea and other waste matters dissolved in
a large quantity of water.[1]

The kidneys are very important organs in helping to
keep the blood pure.   They rid it of the nitrogenous
portion of the waste products which it is always re-

[1] The whole of this excretion is called the **urine**.  It is in reality water,
holding in solution urea and several other salts.  The urine is constantly being
secreted by the kidney.  It is carried to the bladder by two tubes, the bladder
serving as a reservoir.  It collects in the bladder until that organ is nearly full,
when it is emptied by contraction of its walls, aided by the abdominal muscles.

ceiving. This is passed off for the most part in the urine.

About three pints of fluid are daily discharged, on the average, through the kidneys, and a little over one ounce of urea, containing about two hundred and thirty grains of nitrogen. Out of the body, the urea soon changes into carbonic acid and ammonia.

The kidneys also serve to carry off various chemical and mineral substances that are either foreign to the body, or are present in the blood in too large a proportion. Thus, if we should drink a glass of salt water, the kidneys would rid the body of the excess of salt.

If the kidneys are inactive, or fail to excrete the nitrogenous waste matters, the work of many other organs is seriously impaired. The blood is poisoned, and death may result, from the powerful poisons left in the body.

**201. The Health of the Kidneys.** — The kidneys are very busy organs. Their health is a matter of prime importance to every human life. We have already become familiar with the hygiene of two great organs of excretion, — the skin and the lungs. These three sets of organs, then, working together in harmony, like three groups of mechanics doing some difficult work, keep the bodily machinery from getting clogged and choked with waste matters.

If the balance of healthy life is to be maintained, these three great vital processes, or functions, must be carried on or regulated in strict harmony with each other. Thus, if the free action of the skin is interrupted, the kidneys have extra work to do. They make every effort to do the additional work that is thrown upon them ; but, sooner or later, they fail under the bur- den, and become diseased.

Again, a person may overstrain the kidneys by drink- ing any fluid in excessive amount. This is especially true of those who drink enormous quantities of beer. This, in time, results in disease of these organs. In certain diseases, often due to the use of alcohol, the pre- cious albumen is drained off from the blood through the kidneys.

Like the liver, the kidneys are often singled out by well-meaning people as a favorite seat of imaginary disease. If a person gets cold, strains the deep mus- cles of the back, or is otherwise crippled, he is too apt to speak of the "crick" in the back as a "trouble with my kidneys."

The truth of the matter is, that the most serious and insidious diseases of these organs are rarely accom- panied with pain. Hence the popular use of all kinds of nostrums, shrewdly prepared to increase the flow of the urine, and called a "sure cure" for all kidney ail- ments, rarely does any good. Nine times out of ten, all such aches and pains in the back are due to a severe cold or a muscular strain.

**202. Effect of Alcohol upon the Kidneys.** — The kidneys are quick to feel the effects of alcohol. They become diseased in various ways, and often to a fatal extent even by what is called the "moderate" use of alcohol.

Indeed, so powerful is the alcoholic poison on the vital action and structure of the kidneys, that physicians who are best fitted to judge, say that at least three-fourths of the instances of kidney disease are due to the use of alcohol, and to this cause alone.

In the first place, alcohol excites the kidneys to over-action by a chronic dilatation of the blood-vessels, thus irritating the delicate excreting tubes of these organs. Again, it is known that hindering the changes of tissue-waste prevents the proper reduction into urea.

Now, it is generally admitted, that alcohol does, in some measure, hinder the proper changes in the tissue-waste. Hence a large part passes out in a less soluble and more irritating form.

Either from this cause, or from repeated congestion and irritation, alcohol leads at last to widespread disease of these vital organs in a large number of cases. The structure of the kidneys may be so changed from the use of alcoholic drinks, that urea is not properly excreted ; and albumen is drained off instead.

This may give rise to that peculiarly insidious and fatal disease known as "Bright's disease." [1]  It is now generally admitted, that the habit of drinking alcoholic liquors is a frequent cause of this dreaded disease.

The structure of the kidney is so changed, that there is no thoroughfare for the natural amount of urea.  Thus,

[1] So called from the name of the English physician who first described the disease.

" The relation to Bright's disease is not so clearly made out as is assumed by some writers, though I must confess to myself sharing the popular belief that alcohol is one among its most important causes."—ROBERT T. EDES, M.D.

" Whether alcohol or cold be the sharper blade of the shears so apt to cut short life's thread by Bright's disease, I care not greatly, inasmuch as it cannot be questioned that both of them have that tendency." —CHAMBERS's *Diet in Health.*

the ordinary safety-valve is shut off.    The albumen, so
essential to healthy blood, is filtered off through the
diseased organ ; and the waste matters — the ashes —
get into the blood, and poison it. [1]

[1] "As a primary affection, Bright's Disease occurs especially in persons
addicted to intemperance." — AUSTIN FLINT, M.D.

# CHAPTER XIV

## THE THROAT AND VOICE

**203. The Throat.** — The throat is the common passage through which food goes to the stomach, and air to the lungs.

It is shut in and protected by the muscles and bones of the neck. Both the outside of the neck and its interior, or the throat, are exposed to sudden and frequent changes of the weather, which oftentimes lead to diseases of the throat and the lungs. We have already learned something of the food and air passages in the preceding chapters.

The only way to get a proper idea of the throat, is to look into a friend's mouth. First let the person hold his mouth wide open, facing a good light. Hold the tongue down with the handle of a spoon. We are already familiar with the hard palate, soft palate, uvula, and tonsils. Now, on looking directly past these organs, we see the beginning of a passage called the **pharynx,** which is common to the two highways by which air and food are taken into the inside of the body. Look sharp at the top, and we see the air-passage which leads to the nose. The air-tract at the top has two outlets, the mouth and the nose.

Now, if we pull the tongue forcibly forward, a little curved ridge is sometimes seen behind it. This is the **epiglottis,** which, as we already know, is the trap-door

which shuts down, like the lid of a box, over the top of
the air-passage, or windpipe.

The part of the throat directly opposite the line of
vision, as we have just learned, is the back wall of the
pharynx, which, as the gullet, or food-passage, continues
downward. The pharynx runs up to the base of the
skull behind, and ends in a kind of vaulted roof, shaped
something like a crooked forefinger, or the top of an
old-fashioned chaise.

All the parts of this irregular-shaped region, known
as the throat, are covered with a lining-membrane which
secretes a fluid to keep them moist.

**204. Care of the Throat.** — Exposed as it is to
unhealthy and unwholesome air, irritating dusts of the
street and the workshop, it is not strange that the delicate lining of the throat becomes inflamed. The result
is an ailment which is popularly called "sore throat."
Almost every one has at times suffered more or less
from it. The most frequent cause of throat-disease
is the direct action of cold upon the heated body,
especially during profuse perspiration.

Eating hot food, and then drinking ice-water to
cool the parts, is another source of sore throat. We
may overstrain the muscles of the throat in loud talking,
shouting, or by reading aloud too much. All classes of
persons who strain the voice, or misuse it, often suffer
from a severe kind of sore throat, popularly called
"clergyman's sore throat."

Persons subject to throat-disease should take great
pains to wear proper underclothing. Daily baths are
excellent tonics to the skin, and thus serve indirectly

to harden one liable to throat ailments to ordinary changes in the weather. Muffling the neck in scarfs, furs, and wraps is not a good plan, — it only increases the liability to catch cold, — except, perhaps, during the coldest weather, or during unusual exposure to cold.

**205. The Organ of Speech.** — As we have been told in a preceding section, the box-like top of the wind-pipe is called the **larynx,** meaning "top of the wind-pipe." It is composed mainly of cartilage.

The sides of this box are made of two flat pieces of cartilage shaped like a shield.[1] The edges unite in front, and project to form " Adam's apple," which is easily felt, and is plainly to be seen on most lean people, especially spare men. This cartilage shelters the delicate and movable structure within, and shields it from injury without. The epiglottis is attached to the inner and upper part of this cartilage.

Just below is a ring-shaped cartilage.[2] It is broad behind, quite narrow in front, much like a seal-ring. This is easily detected under the skin, a little below the Adam's apple. Two slender, ladle-shaped cartilages[3] are placed on the top of the back part of the cricoid. They work with a ball-and-socket joint, and

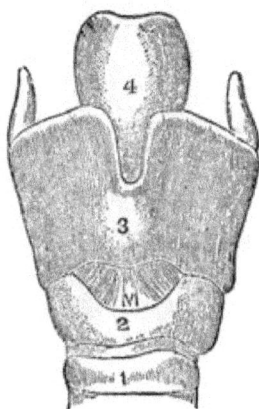

FIG. 75. — A Front View of the Exterior of the Cartilages of the Larynx: 1, Upper Ring of the Windpipe; 2, Cricoid, or Ring-like Cartilage, the Base of the Larynx; 3, Thyroid, or Shield-like Cartilage (the figure 3 is on the Adam's apple); 4, Epiglottis; M, a Membrane uniting the Cricoid and the Thyroid Cartilages.

[1] Thyroid cartilage.    [2] Cricoid cartilage.    [3] Arytenoid cartilages.

have tiny muscles which regulate their movements with the utmost accuracy and regularity. From each of these two cartilages a band of elastic tissue passes forward, and is tied to the inner and front part of the shield-like cartilage towards its lower edge.

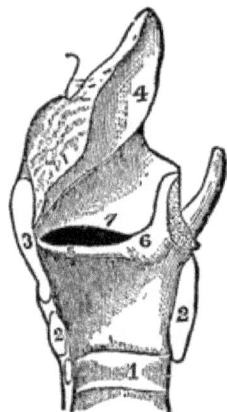

These two bands, called the **vocal cords,** are narrow strips of firm, fibrous material, with a chink-like opening called the **glottis,** meaning the "mouth-piece of a flute." This is the real organ of voice. All the air which goes out or into the lungs must go through the glottis. Muscles and cartilages act to tighten or loosen these cords.

FIG. 76. — Interior View of Right half of the Larynx : 1, Upper Ring of the Wind-pipe; 2, Cricoid Cartilage; 3, Thyroid Cartilage; 4, Epi-glottis; 5, Vocal Band (Vocal Cord); 6, Arytenoid Carti-lage; 7, Ventricular Band (so-called False Vocal Cord).

As the air is forced through the glottis, these bands are set into vibration, and thus sound is produced. When the cords are tightly stretched, and near together, they vibrate more rapidly, and send out a high tone: but when they are less tense, and wider apart, their vibrations are less rapid; and a relatively low tone is the result. During ordinary breathing, the vocal cords are widely separated.

**206. How the Voice is produced.** — If the air be driven out of the lungs by an act of expiration, when the cords are in a state of tension, they vibrate, and produce the sound called the **voice.**

The different musical sounds produced in singing

depend upon the varying degree of tension of the vocal cords. The compass of the voice depends upon the extent to which the variations can take place. A practised singer can, at will, give the requisite tension for the production of any particular note.

The quality of a voice depends upon the structure of the larynx. In women and children [1] the larynx is smaller, and the vocal cords shorter, than in men: consequently their voices have a higher pitch. The longer the cords, and the larger the larynx, the deeper the voice.

FIG. 77. — The Vocal Cords during Inspiration.

Voice may exist without speech, as in many animals. Speech is the voice modified by the throat, teeth, palate, nose, tongue, and lips. In whispering, words are uttered without that vibration which gives vocal sound.

**207. Effect of Alcohol and Tobacco upon the Throat and Voice.** — The peculiar harsh tone to the voice of those given to strong drink is a familiar fact. The reason for it is plain. Alcoholic liquors inflame and irritate the delicate lining of the throat and of the vocal cords. This, after a time, makes the mucous membrane lining thick and rough.

Alcohol weakens the vocal efforts. Hence vocalists, clergymen, and public speakers find that alcoholic drinks impair the voice.

Tobacco tends always to weaken the vocal effort.

---

[1] About the age of fourteen, the vocal organs begin to enlarge rapidly. It causes a "change of voice," as it is commonly called, and is most marked in boys. The voices of children are very much alike in both sexes.

Strong tobacco-smoke affects the vocal cords and to a marked degree injures the voice. In fact, our best teachers of vocal music forbid their pupils to use tea, tobacco, and alcoholic liquors.[1]

Breathing air laden with tobacco-smoke is apt to produce sore throat. The habit of inhaling or swallowing tobacco smoke or breathing it out through the nose, is hurtful to the throat and often results in permanent injury.

Cigarette-smoking also is very irritating to the throat-passages. It makes the throat red and sore.

[1] The world-famous throat doctor, the late Sir Morel Mackenzie, of London, strongly objected to cigarettes. These are his words:—

'Cigarette-smoking is the worst form of tobacco indulgence, from the fact that the very mildness of its action tempts people to smoke nearly all day long, and by inhaling the fumes into their lungs saturate their blood with the poison. It should be borne in mind that there are two bad qualities contained in the fumes of tobacco. One a poisonous nicotine, the other the high temperature of the burning tobacco. Excessive indulgence in the habit is always injurious.'

## CHAPTER XV

### SIMPLE MATTERS OF EVERY-DAY HEALTH. — WHAT TO DO, AND HOW TO DO IT

**208. Accidents and Emergencies.** — Accidents are liable to happen to any one. A friend may cut his leg or foot with a scythe or knife; a child may accidentally swallow some laudanum, or push a bean into his nose or ear; a teamster is brought in with his ears frost-bitten; a small boy falls into the river, and is brought out apparently drowned; some one of our own family may be taken suddenly sick with some contagious disease, or may be suffocated with coal-gas.

All these, and many other things of a like nature, are likely to call for a cool head, a steady hand, and some practical knowledge of what is to be done. These homely details of health may be easily mastered. It is simply the practical application of what we have studied thus far about our bodies.

The dry facts of anatomy and physiology become tenfold more attractive and instructive to us when we thus learn to apply what we have studied to the simple details of daily health.

Let us now try to become familiar with a few of the simplest helps in ordinary accidents, emergencies, and in other similar matters of personal health.

**209. Poultices, how to make them, and when to use them** — There are many occasions when the right kind of a *poultice*, made in the right way, and applied to the right spot, will give a great deal of relief from sudden and severe pains in the chest, abdomen, and elsewhere. The materials are so handy, the relief afforded so great, and the risk of doing harm so little, that even a child should know how to make a poultice, or put on a hot outward application.

Foremost amongst these household remedies is mustard. To make a "mustard paste," as it is called, mix one even tablespoonful of mustard and three or four of fine flour with enough water or vinegar to make the mixture an even paste. Spread it neatly with a table-knife on a piece of old linen, or even cotton cloth, of the required size. Cover the face of the paste with a thin piece of old muslin or linen. Now apply the plaster with its face down.

Some skins are easily blistered with mustard, hence it must be watched. About twenty minutes is long enough to keep it on. If the skin smarts badly, and is quite red, the paste should be removed. Do not put mustard paste on children : their skin is easily blistered with it.

A thicker paste, or a poultice proper, can be made out of flaxseed, oatmeal, rye-meal, ground slippery elm, or bread. A poultice to be of any good, and to hold its heat, should be from half an inch to an inch thick.

The way to make a good poultice is short and simple. Stir the meal slowly into a bowl of boiling water, just as the old-fashioned "hasty-pudding" is made, until a thin and smooth dough is formed. Fold a piece of old linen of the right size in the middle; spread the dough evenly on one-half of the cloth, and cover it with the other. The secret of using a poultice is to apply it hot, and keep it so by frequent changes. Keep the poultice mixture hot on the stove, use duplicate cloths, and have a fresh, hot poultice to exchange in an instant for the cold one. Never let a poultice get cold: it will do more harm than good.

Flannel wrung out in hot water, and cloths heated dry, and applied to painful places, are called *fomentations*. Simply dip the flannel, folded several times into the desired shape, into boiling water; put it into a towel to protect the hands, and wring it out well. Apply it hot as possible, cover it carefully with larger pieces of flannel (oiled silk is the best of all), and use duplicate pieces to make the changes rapidly.

These outward applications are useful to relieve sudden cramps and pains in the abdomen and chest, due to severe colds, injuries, or sprains. Care should be taken to prevent the bedding or clothes from being wet or dampened. A "crick" or "stitch" in the side or back is quite common from overstraining certain muscles, and catching cold. A swollen face, due to an ulcerated tooth or neuralgia; flying pains in various parts of the body; a colicky pain in the abdomen; boils and other sores tending to "point," and very painful, — are a few of the many ailments for which poultices and outward applications will give a temporary relief from suffering.

Rubber bags made to hold very hot water are useful helps in keeping outward applications constantly hot.

### ACCIDENTS AND EMERGENCIES.

**210. What to do for a Fainting Person.** — Fainting is too familiar a sight to need much comment. Very little treatment is necessary. A fainting person should be laid flat at once. Give plenty of fresh air ; and dash cold water, if necessary, on the head and neck. Loosen all tight clothing. It is well enough to hold smelling-salts to the nose. to excite the nerves of sensation.

**211. Epileptic Fits, Convulsions of Children, etc.** — Sufferers from "fits" are more or less common. There is often a peculiar cry, a loss of consciousness, a moment of rigidity, and violent convulsions come on : there is foaming at the mouth, the eyes are rolled up, and the tongue or lips are often bitten.

Such a sufferer should be treated much the same as for fainting-fits. See that the person does not injure himself ; crowd a corner of a folded handkerchief between the teeth, to prevent biting of the lips or tongue. Persons who are subject to such fits should not go out alone very much, and never into crowded or excited gatherings of any kind.

Children occasionally will eat some indigestible substance, and suddenly be taken with convulsions. Give an emetic at once, and an enema of warm soapsuds. Put the child's feet and legs into a bath of hot mustard-water for fifteen minutes.

**212. Asphyxia, or Suffocation.** — When for any reason the proper supply of oxygen is cut off, the tissues rapidly load up with carbonic acid. The blood turns dark, and does not circulate. The healthy red or pink look of the lips and finger-nails becomes a dusky purple.    The person is suffering from a lack of oxygen; that is, from asphyxia, or suffocation, meaning "absence of pulse."

The first and essential thing to do is to give fresh air.    Remove the person to the open air, loosen all tight clothing, dash on cold water, give hot water with a few drops of ammonia in it or hot ginger-tea, and, if necessary, use artificial respiration, as stated in the next section on apparent drowning.

The chief dangers from poisoning by noxious gases come from the fumes of burning coal in the furnace, stove, or range; from "blowing out" gas, or turning it down, and having it blown out by a draught; from the foul air often found in old wells; from the fumes of charcoal and the foul air of mines.

**213. Apparent Drowning.** — Four things have happened to a person nearly dead from drowning.    First, he is unconscious, because he has not had enough oxygen.    Second, he has sucked water into the air-passages, and it must be removed.    Third, he is cold, and needs warmth.    Fourth, his blood is so full of carbonic acid, that the circulation is at the lowest ebb.    ·What must be done?

Remove all tight clothing from the neck, chest, and waist.    Sweep the forefinger, covered with a handkerchief or towel, round through the mouth, to free it from froth and mucus.    Turn the body on the face, raising it a little, with the hands under the hips, to allow any water to run out from the air-passages.    Take only a moment for this.

Lay the person flat upon the back, with a folded

coat, or pad of any kind, to keep the shoulders raised
a little.  Remove all the wet, clinging clothing that is
convenient.  If in a room or sheltered place, strip the
body, and wrap it in blankets, overcoats, etc.

If at hand, use bottles of hot water, hot flats, or bags
of hot sand round the limbs and feet.  Watch the

FIG. 73. — The process of artificial respiration : first step.

tongue : it generally tends to slip back, and to shut off
the air from the glottis.  Wrap a coarse towel round
the tip, and keep it well pulled forward.

The main thing to do is to keep up artificial respiration until the natural breathing comes, or all hope is
lost.  This is the simplest way to do it : The person
lies on the back ; let some one kneel behind the head.
Grasp both arms near the elbows, and sweep them upward above the head until they nearly touch.  Make
a firm pull for a moment.  This tends to fill the lungs
with air by drawing the ribs up, and making the chest-

cavity larger. Now return the arms to the sides of the body until they press hard against the ribs. This tends to force out the air. This makes artificially a complete act of respiration. Repeat this act about fifteen times every minute.

All this may be kept up for several hours. The first

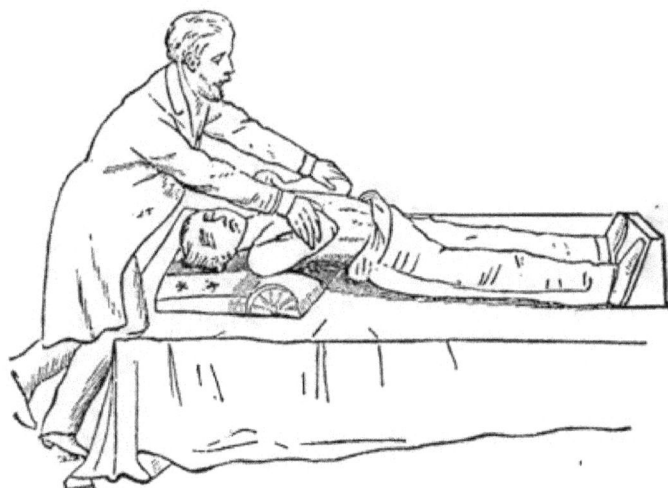

FIG. 79. — The process of artificial respiration : second step.

sign of recovery is often seen in the slight pinkish tinge of the lips or finger-nails. Because the pulse cannot be felt at the wrist is of little account as a sign of death. Life may be present when only the most experienced ear can detect the faintest heart-beat.

When a person can breathe, even a little, he can swallow. Put smelling-salts or hartshorn to the nose. Put one teaspoonful of the aromatic spirits of ammonia, or even of ammonia water, into a half-glass of hot water, and give a few teaspoonfuls of this mixture every three minutes. Meanwhile do not fail to keep up artificial warmth in the most vigorous fashion.

Do not move a person who is just beginning to breathe again from one place to another for some time, except when forced to do so from cold, or any pressing necessity. Above all things, do all that you do in right good earnest, and not in a spasmodic, half-hearted sort of way.

**214. Sunstroke or Heatstroke.** — This severe trouble is caused by an unnatural elevation of the bodily temperature by exposure to the direct rays of the sun, or from the extreme heat of close and confined rooms, as in the cook-rooms and laundries of hotel basements, from overheated workshops, etc. The worst cases of "sunstroke" often happen in places where the sun's rays never penetrate.

There is sudden loss of consciousness, with deep, labored breathing, an intense burning heat of the skin, and a marked absence of sweat. The main thing is to lower the temperature. Strip off the clothing; apply chopped ice, wrapped in flannel to the head. Rub ice over the chest, and place pieces under the armpits and at the sides. If there is no ice, use sheets or cloths wet with cold water.

The body may be stripped, and sprinkled with ice-water from a common watering-pot. If the skin is cold, moist, or clammy, the trouble is due to heat-exhaustion. Give plenty of fresh air, but apply no cold to the body. Apply heat, and give hot drinks, like hot ginger tea. Remember that sunstroke or heatstroke is a dangerous matter. It may be followed by serious and permanent results.    Persons who have once suffered in this way should carefully avoid any risk in the future.

. **215. Broken Bones.** — Send for a doctor at once. Loss of power, pain, and swelling are symptoms of a broken bone, that may be easily recognized.

Broken limbs should always be handled with great care and tenderness. If the accident happens in the woods, the limb should be bound with handkerchiefs,

**How to carry an Injured Person.** — "If injured persons have to be moved from one place to another, it is worth while to know how to do it with the greatest ease and safety to them. If a door or shutter or settee is at hand, any of these will make a good litter, with a blanket or shawls or coats for pillows. In lifting a person upon a stretcher, it should be laid with its foot at his head, so that both are in the same straight line. Then one or two persons should stand on each side of him, and, raising him from the ground, slip him upon the stretcher. This can be done smoothly and gently; whereas, if a stretcher is laid alongside of an injured person, some of those who lift him will have to step backwards over it, and in doing so are very apt to stumble. If a limb is crushed or broken, it may be laid upon a pillow, with bandages tied round the whole, so as to keep it from slipping about. Where an injured person can walk, he can get much help by putting his arms over the shoulders and round the necks of two others. In case of an injury where walking is impossible, and lying down is not absolutely necessary, an injured person may be seated on a chair, and carried; or he may sit upon a board or fence-rail, the ends of which are carried by two men, around whose necks he should place his arms, so as to steady himself; or two men may carry him seated on their interlocked hands, in the way known to children as 'Lady to London.' To do this, each of two persons standing face to face should grasp his right fore-arm with his left hand (its back uppermost), then he should grasp his companion's free left fore-arm with his own free right hand (also with its back uppermost). When no litter can be gotten, the body may be supported by a man on each side, with their arms placed behind his chest and under his hips. In carrying an injured person upon a litter, or what serves for one, the bearers ought not to keep step; but, when they are not using a litter they should keep step."— DULLES'S *Accidents and Emergencies.*

suspenders, or strips of clothing, to a piece of board,
pasteboard, or bark, padded with moss or grass, which
will do well enough for a tem-
porary splint.  Always put a
broken arm into a sling after
the splints are on.

Never move the injured
person until the limb is made
safe from further injuries by
putting it into splints.  If
you do not need to move
the person, keep the limb in
a natural, easy position, until
the doctor comes.

FIG. 30. — Showing how a temporary
splint may be put on to a broken leg.

FIG. 31. — Showing how a temporary
splint and a sling may be put on to a
broken arm.

Keep the patient warm.   Do not give a drop of alcoholic liquor.

Remember that this treatment for broken bones is

only to enable the patient to be moved without further injury.    A surgeon is required to set the broken limb.

216. **Burns or Scalds.** — Burns or scalds are danger-ous in proportion to their extent and depth.    A child may have one of his fingers burned off with less danger to life than an extensive scald of his back.

After severe burns or scalds, remove the clothing with the greatest care.    Do not pull, but carefully cut and coax, the clothes away from the burned places, taking some time for it.    Save the skin unbroken if possible, taking care not to break the blisters.    The secret of treatment is to avoid chafing, and to keep out the air.

If the burn is slight, put on pads of soft linen soaked in a strong solution of baking-powder and water, one heaping tablespoonful to a cupful of water.    It may be put on dry, and kept well covered with cotton-wool or bandages.    "Carron oil" is one of the best applica-tions.    It is simply half linseed-oil and half lime-water shaken together.    Soak strips of old linen in this, and gently apply.    Simple cold water is better than flour, cotton-batting, and other things which are apt to stick, and make an after-examination very painful.    Re-member this : —

A deep or extensive burn or scald should always have good medical attendance.

217. **Frost-bite.** — The ears, toes, nose, and fingers are occasionally frozen or frost-bitten.    No warm air, warm water, or fire is to be allowed near the frozen parts until the natural temperature is nearly restored. Great care must always be taken in the after-treatment,

lest serious results should follow.  Rub the affected part gently with snow or snow-water in a cold room.  The circulation should be restored very slowly.  Hot milk with cayenne pepper, and hot ginger-tea should be freely given.

218.  **Foreign Bodies in the Throat.**—Bits of food and other small objects sometimes get lodged in the throat, and are easily got out by the forefinger, by sharp slaps on the back, or thrown out by vomiting.  If it is a sliver from a toothpick, match, or fish-bone, it is no easy matter to remove it; for it generally sticks into the lining of the passage.  Holding a child up by the heels, with the head hanging down, and giving a few vigorous shakings, will often dislodge a foreign body when simpler means fail.

If the object has actually passed into the windpipe, and followed by sudden fits of spasmodic coughing, with a dusky hue to the face and fingers, there is great danger of life.  Surgical help must be called without delay.

If a foreign body, like coins, pencils, keys, fruit-stones, etc., is swallowed, it is not wise to give a physic.  Give plenty of hard-boiled eggs, cheese, and butter-crackers, so that the substance may be passed off in the natural way, in a bulky stool.

219.  **Foreign Bodies in the Nose.** — Children are apt to push marbles, beans, peas, fruit-stones, buttons, and other small objects, into the nose.  Sometimes we can get the child to help by blowing the nose hard.  At other times, a sharp blow between the shoulders will cause the substance to fall out.

Do not waste time, — especially if it is a pea or bean, which is apt to swell with the warmth and moisture, — but call in medical help at once if you do not meet with success.

**220. Foreign Bodies in the Ear.**— It is a much more delicate matter to get foreign bodies out of the ear than from the nose.   The simplest thing is to syringe in a little warm water, which will often wash out the substance.   If live insects get into the ear, drop in a little sweet-oil, salt and water, or even molasses.

If the tip of the ear is pulled up gently, the liquid will flow in more readily.   If a light is held close to the outside ear, the insect may be coaxed to crawl out towards the outer opening of the ear, being attracted by the bright flame.

**221. Foreign Bodies in the Eye.** — Cinders, particles of dust, and other small substances, often get into the eye, and cause much pain.   Do not rub the eye, however much you may feel like it.   It will only make bad matters worse.   Often the copious flow of tears will wash the substance away.   It is sometimes seen, and removed simply by the twisted corner of a handkerchief carefully used.   If it is not removed, or even found, in this way, the upper lid must be turned back.

This is the way to do it : Seize the lashes between the thumb and forefinger, and draw the edge of the lid away from the eyeball.   Now, telling the patient to look down, press a slender lead-pencil or penholder against the lid, parallel to and above the edge, and then pull the edge up, and turn it over the pencil by means of the lashes.

The eye is now easily examined, and usually the foreign body is easily seen and removed. Do not increase the trouble by "poking" at the eye after you fail, but get at once skilled help. After the substance has been removed, bathe the eye for a time with hot water. If lime gets into the eye, use large quantities of vinegar or lemon-juice and water, one teaspoonful to a teacupful of water.

**How to remove a Foreign Substance from the Eye.** — " Lay your finger on the cheek and draw the lower lid gently down, while the person looks as much upward as possible, and we shall see about the whole extent of the lower portion of the conjunctiva; and thus if any foreign substance is there, it will be readily detected, and easily wiped away with a folded soft rag or handkerchief. Both lids have a piece of cartilage in them to stiffen them, like pasteboard, and keep them fitting close to the eyeball.

" The upper portion of this conjunctival sac can only be seen by turning over the upper lid. The way to do this is to let the person look down with the eyes closed. Taking hold of the lashes with one hand, and applying a pencil, or some small, round, smooth object, over the lid above the globe, we lift the lashes out and up, warning the person to still keep looking down. The lid will suddenly turn over with a little spring from the bending of the cartilage.

" In this way nearly the whole of the conjunctival sac will be exposed, and any foreign body wiped away, as above described. But suppose no friend or oculist is by us to do this. The next best thing is to take hold of the lashes of the upper lid, and draw it forward and downward over the lower one, blowing the nose violently with the other hand at the same time.

" If the foreign substance is on the cornea, take a strip of paper not stiffer than ordinary writing-paper, about a quarter of an inch wide, and roll it up as if you were going to make a candle-lighter. Look at the lower end, and you will see it comes to a point. With this point now you may safely attempt to remove any foreign substance from the cornea. The tears which will flow soften the paper, and prevent injury to the delicate covering membrane of the cornea."— Dr. B. Joy Jeffries.

### HEMORRHAGE — BLEEDING.

**222. Injuries to the Blood-vessels.** — It is very important to know the difference between the bleeding of an artery and that of a vein. If an artery bleeds, the blood spurts in a stream, or leaps in jets, comes very rapidly, and is of a bright scarlet color. If a vein bleeds, the blood slowly oozes out, or flows in a steady stream, and is of a dark and purple color.

Bleeding from an artery is a dangerous matter in proportion to the size of the vessel, and life itself may be speedily lost ; while hemorrhage from the veins is rarely a serious injury, and generally stops of itself, aided, if need be, by hot water, deep pressure, some form of iron styptic, or even strong alum-water. When an artery is bleeding, always remember to make deep pressure between the wound and the heart. In all cases where common sense would suggest it, send at once for the doctor.

Meanwhile there is something to do. Keep cool, remember the simplest facts of the anatomy we have just studied, and do not be afraid to act at once. A resolute grip in the right place with firm fingers will do well enough, until a twisted handkerchief, stout cord, shoestring, or suspender is ready to take its place. If the artery is of some size, make a knot in whatever is used, and bring the pressure of the knot to bear over the artery. If the flow of blood does not stop, change the pressure until the right spot is found.

Sometimes it will do to seize a handful of dry earth, and crowd it down into the bleeding wound, with a firm pressure. Strips of an old handkerchief, underclothing, or cotton wadding, might be stuffed into the wound, keeping up the pressure all the time.

Let us try to keep in mind the principal places to apply pressure when arteries are injured and bleeding.

If in the *finger*, grasp it with the thumb and forefinger, and pinch it firmly on each side : if in the *hand*, press on the bleed-ing spot, or press with the thumb just above, and in front of, the wrist.

For injuries *below the elbow*, grasp the upper part of the arm with the hands, and squeeze hard. The main artery runs in

FIG. 82. — Showing how a handker-chief and a stick may be applied to the arm to stop bleeding.

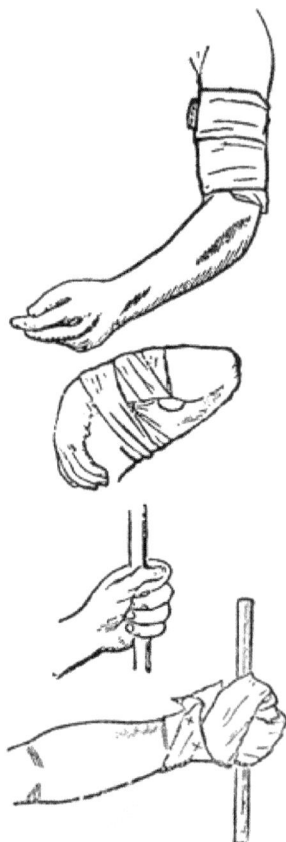

FIG. 83.— Showing how a bandage may be used to stop bleeding from an artery in the arm.

the middle line of the bend of the elbow. Tie the knotted cord here, and bend the fore-arm so as to press hard against the knot.

For the *upper arm*, press with the fingers against the

bone on the inner side, and just on the edge of the swell of the biceps muscle. Now we are ready for the knotted cord. Take a stout stick of wood, about a foot long, and twist the cord hard with it, bringing the knot firmly over the artery.

For the *foot* or *leg*, pressure as before, in the hollow behind the knee, just above the calf of the leg. Bend the thigh towards the abdomen, and bring the leg up against the thigh, with the knot in the bend of the knee. A knife may glance, and cut the great femoral artery. Remember it runs deep in the inside of the thigh, but is exposed when it comes out of the body below the groin. Press in the hollow just below the groin, about two-thirds of the way from the hip-bone to the middle line of the body. Double the thigh against the abdomen, and bend the leg on to the thigh, with the knotted cord in position.

**223. Bleeding from the Stomach and Lungs.**— Blood that comes from the lungs is bright red, frothy, or "soapy." There is rarely much; it usually follows coughing, feels warm, and has a salty taste.

Bleeding from the lungs is a grave symptom. Perfect rest and quiet must be insisted upon. Bits of ice should be eaten freely. Loosen the clothing, keep the shoulders well raised, and the body in a reclining position.

Blood from the stomach is not frothy, has a sour taste, and is usually dark colored, looking some like coffee-grounds. It is larger in quantity, and apt to be mixed with food. Do the same as before, except keep the person flat on the back instead of in a reclining position.

**224. Bruises, Cut and Torn Wounds. —**A bruise is a familiar sight. It is a wound of the soft tissues, caused by blows. It is more or less painful, followed by discoloration due to the escape of blood under the skin. A black eye, and a lip or finger hurt by a base-ball, are familiar examples of this sort of injury. Soak the injured part in ice-cold-water cloths at first, and, after the pain is easier, soak the parts in water hot as can be borne. If the cuts are small, clean the parts, bring the edges together, and stick them with plaster.

When wounds are made with ragged edges, such as those made by broken glass and splinters, more skill is called for. Every bit of foreign substance must be got rid of first. Wash the parts clean with warm water, bring the torn edges together, and hold them in place with strips of plaster. Do not cover such a hurt all over with plaster, but leave plenty of room for the escape of matter.

Wounds made from toy-pistols, percussion-caps, and rusty nails, if neglected, often lead to serious results from blood-poisoning. Cloths wrung out in cold water may be enough, but often a hot flaxseed poultice is needed for several days. Keep such wounds perfectly clean.

**225. The Pernicious Use of Alcoholic Liquors in Accidents and Emergencies. —** When an accident or sudden illness occurs, people, otherwise very sensible, are often beset with the idea that brandy or whiskey is the great cure-all for all emergencies. They do not know what to do, but they want to do something.

And so they are prone to give the sick or injured person a liberal dose of strong drink. It is so common a practice that it has passed into common speech as a joke.

Remember this : on such occasions, alcoholic liquors are not only unnecessary, but actually harmful. They injure the patient, mislead the doctor, and interfere also with the proper treatment of the case. Any simple, hot drink will answer every purpose.

### CARE OF THE SICK-ROOM.

226. **Hints for the Sick-room.** — The sick-room should be the lightest and most pleasant room in the house. Some one of the family may be taken sick in the attic, or some inconvenient room. If possible, and especially if there is a prospect of a long illness, at once get a more suitable room ready.

Take away all extra carpets, upholstered furniture, heavy curtains, etc. A clean floor, with a few rugs to deaden the footsteps, is much better than a woollen carpet. Carpets, extra clothing, etc., only absorb the impurities, and help keep the room foul.

Let the room be accessible to sunlight and fresh air It is generally best to shade the room somewhat in certain diseases, yet take the time to let in all the sunlight possible consistent with comfort. Sunlight and fresh air are often more efficient helps than drugs. Besides, they cost nothing but a little painstaking and common sense. Pains must be taken to protect the patient from any noise which may disturb him, such as the noise of passing steam and horse cars, heavy teams, and playing children.

Give the sick-room plenty of sweet, fresh air. With a little pains, any room may be supplied with pure air. If you cannot do anything else, cover the sick person all over with extra bed-clothes, open the windows and doors, and fan out the bad air. Be sure and avoid draughts of cold air. Have a thermometer, and keep the temperature as the doctor directs.

Do not allow a kerosene-light, with its flame turned down, to burn through the night. A close room with such odor for a whole night is enough to make a well person sick. If there is no gas, either use the lamp as usual, and put it, carefully shaded, in an adjoining room, or, better still, use a sperm candle for a night-light.

Keep a sick-room neat and trim. Remove at once all offensive matters. Never allow such things to remain for a moment in the room. In many diseases, especially scarlet-fever, diphtheria, putrid throat, consumption, etc., use pieces of old linen instead of handkerchiefs, and burn them as soon as they are used. Carelessness or ignorance in this matter often spreads contagious disease.

Change the clothes of the bed and of the patient quite often. Do not let such clothing be put away in a closet with others. Put them to soak at once in boiling water, with some disinfectant added if necessary. Do not make a great show of bottles of medicines, spoons, glasses, etc., carefully spread out on the bureau or table. Keep all such things, except such as are absolutely necessary, in an adjoining room.

To a patient not used to sickness, a formidable array

of drugs and sick-apparatus is apt to be discouraging. Some simple thing, like an orange, a tiny bouquet of favorite flowers, and one or two playthings, may take their place.

Never get behind the door, in a corner, or in an adjoining room, and *whisper*. It will fret a well person, to say nothing of its hurtful effects upon a sufferer whose nerves may be sensitive to the faintest noise.

Whatever must be said, say it openly and aloud. How often a sudden turn in bed, or a quick glance of inquiry, shows that whispering is doing harm!

If the patient is in his right mind, answer his questions plainly and squarely. It may not be best to tell *all* the truth, but nothing is gained trying to avoid a straightforward reply.

Do not deceive sick people. Tell what is proper or safe to be told, promptly and plainly. If a physician is employed, carry out his orders to the very letter, as long as he visits you. Make a note of his directions on a slip of paper. Make a brief record of exactly what you do, the precise time of giving medicines, etc.

This should always be done in serious cases, and by night-watchers. Then there is no guess-work. You have the record before you in black and white. All such things are valuable helps to the doctor.

Above all, let there be cool, wise heads, willing hands, loving hearts, and a deal of common sense on the part of the helpers in the sick-room; and a thankful submission, and a reasonable patience to endure, in the sick person.

## POISONS AND THEIR ANTIDOTES.

**227. Poisons in General.** — Poisons of various kinds are quite generally used in the trades, and kept about the house and premises as medicines, as disinfectants, for killing insects and animals, and for many other purposes. People are often careless about them, and leave them almost anywhere, wrapped in a piece of paper, or in some unlabelled bottle, even in the cupboard, or on a shelf about the shed or stable.

Children either mistake them, or are urged by some playmate to swallow them. All of us are too apt to seize a bottle or package hastily, and, "before we think," have swallowed a dose of some poison. The many fatal accidents due to drinking carbolic acid by mistake is a familiar example of how stupid or careless people may be.

All poisons should always be put in bottles carefully labelled, and the word "POISON" should be plainly printed in large letters directly across the label. Fasten the cork firmly to the bottle by wire picture-cord, or copper wire, twisted into a knot at the top. This wire would certainly prevent a person from mistaking, in the dark, carbolic acid, oxalic acid, etc., for some favorite medicine.

Poison should never be kept in the same place with medicines or other bottled preparations used in the household. Put them in some secure place, and under lock and key.

Another very simple rule is, never to use the contents of any package or bottle unless we know exactly what it is. Do not guess at it, or take any chances, but destroy it at once.

Poisons are often taken when medical help, especially in the country, cannot be had at short notice. Poisons do their work rapidly. Something must be done, and that at once and in earnest. The stomach must be emptied as speedily as possible. Make a quart of warm soapsuds. Force the sufferer to gulp it down, a cupful at a time. Run the finger "down the throat," and hasten the vomiting.

A good emetic is made by putting a heaping tablespoonful of ground mustard into a pint of water. Drink a cupful every ten minutes until vomiting is produced. Stir up a handful of powdered alum in a cupful of molasses, and swallow this, a tablespoonful every ten minutes.

Be in right good earnest about it, and do not waste time to see if the poisoned person likes such heroic treatment. Vomiting will not do any harm, and the poison may destroy life in a few minutes.

**228. Different Kinds of Poison.** — For convenience, let us arrange the most common poisons into different classes.

Some are **acids,** like the *oil of vitriol;* others are **alkalies,** like *lye.*

Some are irritant **mineral poisons,** like *arsenic* or *sugar of lead;* while others are **vegetable poisons,** like *aconite* and *Jamestown weed.*

We can easily remember the general plan of treatment for each special class of the more common poisons.

**229. Acid Poisons.** — Sulphuric acid (oil of vitriol), nitric acid (*aqua fortis*), and muriatic acid (spirits of salt) are in common use in certain workshops, and occasionally used in the household. These are caustic mineral acids, that rapidly burn and destroy the living tissues.

*Give an alkali.* Swallow large quantities of strong soapsuds, chalk, tooth-powder, soda or saleratus-water, magnesia, or lime-water. Scrape off the whiting from the wall, or dig out a piece of plaster. Put them into large quantities of water, and swallow the mixture. Follow this treatment with some mild, soothing tea made of flaxseed or Irish moss.

**Oxalic acid** is often mistaken for granulated sugar or Epsom salts. For an antidote, use chalk, whitewash, plaster, etc., as before.

**Carbolic acid** in solution is very commonly used about the house. It is a highly dangerous poison, and generally fatal. Provoke vomiting by giving large quantities of soapsuds and sweet-oil mixed together. Follow with large draughts of oil or milk. Give very large doses of Epsom salts.

**230. Alkaline Poisons.** — The common alkalies taken as poisons are **ammonia,** or hartshorn, **potash,** and **soda,** usually dissolved, and often in the form of **lye.** Horse-liniments and other liniments generally contain ammonia, and are often taken by mistake. Alkalies burn the lining-membranes rapidly and severely.

*Give acids.* Drink vinegar freely. Lemon-juice may be used. Take large quantities of sweet-oil, linseed-oil, and castor-oil.

**231. Metallic Poisons.** — **Arsenic** is a white, sweetish powder, used to kill rats. It is occasionally taken by mistake. **Paris green** is a form of arsenic used by farmers.   Arsenic is also found in ratsbane and fly-powder.

Provoke vomiting at once.   Take large quantities of milk, the whites of eggs, flour and water, or oil and lime-water.   A good antidote for arsenic is dialyzed iron, which can be bought at any drug-store.   It should be given freely, in tablespoonful doses, followed by a strong solution of salt and water.

In **sugar of lead** poisoning, provoke vomiting, and give Epsom salts.   In **copper** poisoning by "blue vitriol" and "verdigris," give milk or white of eggs, followed by flaxseed-tea.

Children sometimes eat the **phosphorus** from matches. This poison acts slowly, and there is time enough to get medical help.   Give plenty of magnesia, chalk, or whiting, but no oil.

**232. Narcotic Poisons.** — The various forms of **opium** are often taken by mistake, or in an overdose. The narcotic effects of the following preparations are due to opium ; laudanum, paregoric, morphine, Dover's powder, Godfrey's cordial, McMunn's elixir, most cholera-mixtures, and many of the so-called " soothing-sirups " and " drops."

Brisk emetics must be given until they act thoroughly. Give plenty of hot, strong coffee without milk or sugar. Keep the patient awake, and do not allow him to fall into a deep sleep.   Dash cold water over the head and shoulders, and slap the skin briskly with wet towels or with a slipper.   Medical help must be called at once.

Among other narcotic poisons occasionally taken without a knowledge of their poisonous character, especially by children are **aconite, hemlock, deadly night-shade, Jamestown weed, monkshood,** and **toadstools.** Use essentially the same treatment as in opium-poisoning.

**233. Other Poisons.** — Alcoholic liquors are sometimes taken in sufficient quantities to endanger life **at** once.

Give a brisk emetic, and stimulate with **frequent** doses of a mixture **made of** a teaspoonful of the aromatic spirits of ammonia, **or even** ammonia-water, added to a glass of hot water.

Certain kinds **of fish act sometimes as poisons to certain** people, **such as** eels, crabs, mussels, etc. Fresh pork and other fresh meats occasionally **act as a** mild poison. **Canned goods of a cheap grade** sometimes act in the same way. Induce vomiting with some **brisk emetic, and afterwards give a dose of castor-oil.**

For the **sting of** hornets, wasps, bees, etc., apply a pad of cloth **soaked** in a solution of **carbolic** acid, ammonia, or spirits of camphor **Crowd the barrel of a** watch-key deep **down on** the hurt, to remove the **sting of** the insect.

### DISEASES THAT SPREAD, AND DISINFECTANTS.

**234 How the Air may be Poisoned.** — We have learned in the chapter on Respiration (chap. ix.), that the air may be poisoned by the products **of** respiration, and other useless or injurious matters thrown off from *healthy* human bodies. We have shown the baneful effects of long-continued exposure **in** an atmosphere laden with these impurities.

Now, in like manner, the **air** may be poisoned at cer

tain times and places by the products of respiration and
bodily emanations of *diseased* persons.    Thus, in certain diseases, called *contagious*, organic matters are
thrown off from the persons of the sick, which tend to
reproduce themselves in some way in the bodies of
other persons.

Such diseases as small-pox and scarlet-fever are examples of severe and contagious diseases ; while whoop-
ing-cough, measles, and mumps are contagious, but less
harmful.

Again, the air may be poisoned with the foul gases
arising from the rapidly decomposing contents of cesspools, sewers, and privy-vaults.    This poisoned air is
popularly called sewer-gas.

The living particles, or "germs," of such diseases as
typhoid-fever and dysentery, are believed to be contained, or developed, in the discharges, both from the
stomach and bowels of persons suffering from these
diseases.    Diphtheria and typhoid-fever are believed to
be due, oftentimes, to the decomposition of the contents
of cesspools and sewers.

Finally, the air may be poisoned by the decomposition of organic matters in the ground, and drawn into
the house in various ways.    Thus, there seems to be a
connection between malarial fever and bad drainage.
Certain low, damp soils are especially productive of
consumption.

**235. Disinfection.** — With our present knowledge, it
is not possible to get rid of the germs of disease after
they are once fastened in the body.    We are able, however, to a certain extent, to destroy them after they

leave the body through the emanations and excreta of diseased persons.

This destruction of the poisons of infectious and contagious diseases is called *disinfection*, and the means used are called **disinfectants.**

We must remember that disinfection cannot compensate for the want of cleanliness, nor of ventilation. Those things which destroy bad smells are not necessarily disinfectants, and disinfectants do not necessarily have an odor.

**236. What Disinfectants to use.** — It is our aim to speak only of those which can be relied upon, and can be easily bought in sufficient quantities, even in our smaller villages, at little cost. There are many kinds of patented disinfectants, bearing some high-sounding name, and sold at a high price, which are of no more practical worth than sulphur and copperas, properly used.

It is of the utmost importance, therefore, to know the comparative value of disinfectants, and the most effective methods of using them. It might be a serious mistake, to confound *deodorizers* — which simply kill the odor of disease — with disinfectants, which are believed to kill the germs of disease; and equally so to employ a powerful disinfectant in an ineffective way.

Thus, carbolic acid, so much used in the sick-room, disinfects what it touches, but does not disinfect the air. It is a somewhat costly preparation, and acts more effectively in combination with white vitriol and copperas. Simply to make outbuildings or a sick-room smell strong of carbolic acid, does not disinfect them.

The following disinfectants are the most useful : —

(1) Roll-sulphur or brimstone, for fumigation ;

(2) Sulphate of iron, or copperas, — also called green vitriol, — used in the proportion of one and one-half pound to the gallon of water, for sewers, cesspools, etc. ;

(3) Sulphate of zinc, — called white vitriol, — and common salt, dissolving about four tablespoonfuls of the zinc, and two of the salt, to one gallon of water, for clothing, bed-linen, etc.

**237. How to use Disinfectants.** — The following practical suggestions for the use of disinfectants and the management of contagious diseases, were prepared for the National Board of Health by a committee representing some of the ablest physicians and sanitarians in the country.

1. *In the Sick-room.* — The most available agents are fresh air and cleanliness. The clothing, towels, bed-linen, etc., should at once, on removal from the patient, be placed in a pail or tub of the zinc solution, boiling-hot if possible, before removal from the room.

All discharges should either be received in vessels containing copperas solution, or, when this is impracticable, should be immediately covered with copperas solution. All vessels used about the patient should be cleansed with the same solution.

Unnecessary furniture, especially that which is stuffed, carpets and hangings, when possible, should be removed from the room at the outset : otherwise, they should remain for subsequent fumigation and treatment.

2. *Fumigation.* — Fumigation with sulphur[1] is the only practicable method for disinfecting the house. For this purpose, the rooms to be disinfected must be vacated. Heavy clothing, blankets, bedding, and other articles which cannot be treated with zinc solution, should be opened and exposed during fumigation, as directed below.

Close the rooms as tightly as possible; place the sulphur in iron pans supported upon bricks; set it on fire by hot coals, or with the aid of a spoonful of alcohol, and allow the room to remain closed for twenty-four hours. For a room about ten feet square, at least two pounds of sulphur should be used; for larger rooms, proportionally increased quantities.

3. *Premises.* — Cellars, yards, stables, gutters, privies, cesspools, water-closets, drains, sewers, etc., should be frequently and liberally treated with copperas solution. The copperas solution[2] is easily prepared by hanging a basket containing about sixty pounds of copperas in a barrel of water.

4. *Body and Bed Clothing, etc.* — It is *best* to burn all articles which have been in contact with persons sick with contagious or infectious diseases. Articles too valuable to be destroyed should be treated as follows : —

Cotton, linen, flannels, blankets, etc., should be treated with boiling-hot zinc solution, introducing piece by piece, securing thorough wetting, and boiling for at least half an hour.

[1] The burning of sulphur produces an irrespirable gas. The person who lights the sulphur must, therefore, immediately leave the room, and, after the lapse of the proper time, must hold his breath as he enters the room to open the windows, and let out the gas.

After fumigation, plastered walls should be whitewashed, the woodwork well scrubbed with carbolic soap, and painted portions repainted.

[2] Put copperas in a pail of water in such quantity that some may constantly remain undissolved at the bottom. This makes a saturated solution. To every privy or water-closet use about one pint of the solution night and morning.

Heavy woollen clothing, silks, furs, stuffed bed-covers, beds, and other articles which cannot be treated with the zinc solution, should be hung in the room during fumigation, pockets being turned inside out, and the whole garment thoroughly exposed.    Afterward they should be hung in the open air, beaten, and shaken.

Pillows, beds, stuffed mattresses, upholstered furniture, etc., should be cut open, the contents spread out, and thoroughly fumigated.[1]    Carpets are best fumigated on the floor, but should afterward be removed to the open air, and thoroughly beaten.

---

[1] The cutting open of stuffed articles may seem unnecessary, but it is not.  The poison of contagious diseases may cling to such stuffs with great tenacity for years, and must be destroyed before they are fit to be used again.

" Contagious diseases are often caught at the funerals of those who have died of them ; and the sanitary code of New York City forbids a public funeral of any person who has died of small-pox, diphtheria, scarlet-fever, yellow-fever, or Asiatic cholera.  It is better to limit the attendance at such funerals to as few as possible." — DR. TRACY'S *Handbook of Sanitary Information.*

# CHAPTER XVI

## PRACTICAL EXPERIMENTS

**238. Why we should make Experiments.** — To get a thorough mastery of the elementary principles of anatomy, physiology, and hygiene, it is not enough simply to study the text.

Just as in chemistry and physics, a series of **practical experiments** is made, both by the teacher and the pupils, so should "physiology," as it is popularly called, be illustrated on the same general principles. Geology would certainly be a dry study without suitable specimens to illustrate the text-book.

So, in the study of physiology, it is almost impossible to get an intelligent idea of the text, and that practical knowledge which is so desirable, without the aid of specimens and experiments.

Whatever we see with our eyes, feel with our fingers, and do with our hands, in the matter of experiment and illustration, however simple and homely, is of far more worth than merely studying the printed page. It is more like hard work, to be sure ; but laziness is a poor excuse in this or in any other branch of science.

**239. Importance of making Experiments.** — It is plain that any series of experiments arranged for use in our common schools must be somewhat crude. We must take many things for granted. The observation and experience of medical men, and the experiments of the physiologist in his laboratory, must be relied upon for important data not otherwise easily obtained by young students.

Because we cannot make our experiments with such accuracy and detail as in other branches of science, it by no means follows that we cannot use experiments in the common schools. The simplest experiments become, in the hands of the enthusiastic and skilful teacher, of the greatest value and interest. Pupils soon gain a far better knowledge, and keep up a livelier interest in the subject, if, as we have just said, they see with their own eyes, and handle with their own hands, that which serves to illustrate the subject.

This method of instruction rivets the attention, and keeps alive the interest, of the young pupil; in fact, it is the true method of cultivating a scientific habit of study. Hence the following experiments, however simple and rude the apparatus, become important helps towards gaining a more thorough knowledge of this important branch of study.

**240. General Plan of Study.** — We should depend for this part of our study upon five sets of illustrations : —

(1) **The Skeleton and Manikin.**

(2) **Experiments on the Person.**

(3) **Study of Physiological Charts.**

(4) **Blackboard Diagrams and Sketches.**

(5) **A systematic series of simple Experiments performed by the Teacher and the Pupils.**

**241. Study of a Skeleton or Manikin.** — By all means secure a skeleton and manikin, by loan or purchase, if possible. It is better to have some of the separate bones than none at all.

**242. Experiments on the Person.** — We have suggested in the following pages many ways in which our own person, or that of a friend, may be utilized for the

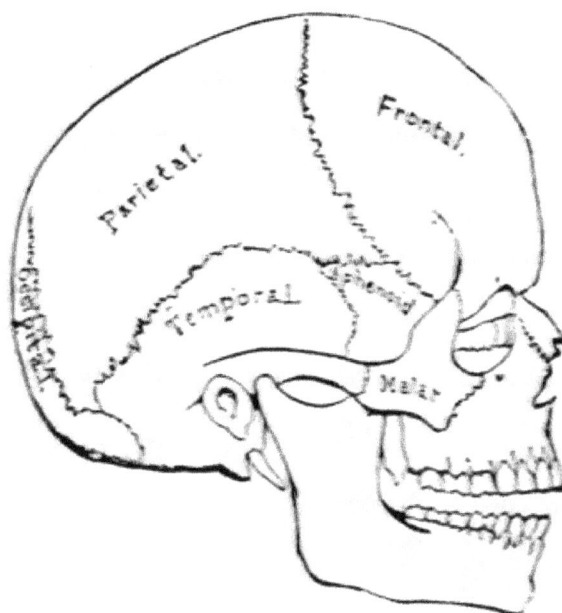

Fig. 54. — BLACKBOARD SKETCH. Right side of the Skull

purpose of experiment and observation. The success of this part of the work will depend upon the skill, tact, and common sense of both teacher and pupils. How much time and effort it is best to devote to this part of the study depends upon the age and capabilities of the class. There is no better way to get an accurate and thorough knowledge of the more essential and practical points of anatomy.

**243. Study of Charts.** — A set of physiological charts is very desirable, and should be secured for every-day reference in the schoolroom.

**244. Blackboard Diagrams and Sketches.** — The pupil should be trained to make and to copy sketches and diagrams in his note-book, and repeat them at stated intervals on the blackboard. Such sketches may be somewhat rude, and there will be a great contrast in the work of different pupils; but this is of little account.

The teacher is advised to make for himself a series of diagrams, on a suitable scale for the blackboard, on sheets of manila paper, and have them bound, and hung up for easy reference in the schoolroom, somewhat in the same way that reading-charts are now made. These sketches may be used to illustrate the text, and then copied by the pupil in his note-book or on the blackboard.

**245. A Series of Simple Experiments.** — By all means, depend upon a systematic series of rude, homely experiments, such as have been suggested in this chapter. They are simply *suggestive.* Do not merely copy them, but contrive others equally good.

In most schools, the cost is an important matter. Accordingly, we have limited most of the experiments to such as can be done with cheap, homely, and simple apparatus, most of which can be picked up about the house, the market, and the village store.

All these experiments should be performed both by the teacher and each member of the class. It is not enough for the student to sit still, and see the teacher do them. Pupils should be encouraged to get together the material, and repeat some of the experiments on Saturday, or during a holiday. Clubs of three or four friends will find it both profitable and interesting to work together on a spare evening.

**246. Hints and Helps.** — When specimens of bones, joints, dissections, etc., are carried into the school-room, the utmost pains must be taken to keep every-thing neat and clean. Use large plates, platters, saucers, tissue-paper, with plenty of pins, needles, clean towels, and napkins. Cover every part except what it is necessary to show. Keep everything covered until the proper time comes.

Every little detail must be arranged before the reci-tation begins. If the fingers get soiled, remove all traces at once. Strict attention to all these little matters may make the difference between success and failure.

If the school does not own a *microscope*, secure the loan of one, if possible, for a few weeks. No one ever forgets the first look at a drop of blood, or the circula-tion of blood in a frog's foot, as revealed by the micro-scope.

Glass slides ready for the microscope, and illustrating every part of the body, can be bought or ordered in any large city, of dealers in microscopic or scientific apparatus.

## EXPERIMENTS ON THE BONES

### [CHAPTER II. PAGE 13.]

EXPERIMENT 1. *To show the Gross Structure of Bone.* —
Saw lengthwise in two the bones of a sheep's leg, or a calf's
leg, including the knee-joint end. Save one-half. Boil,
scrape, and carefully clean the other half. Note in the boiled
half the compact and spongy parts, shaft, etc.

Note on the other half, after trimming off the flesh, the
pinkish-white look of the bone, the marrow and its tiny specks

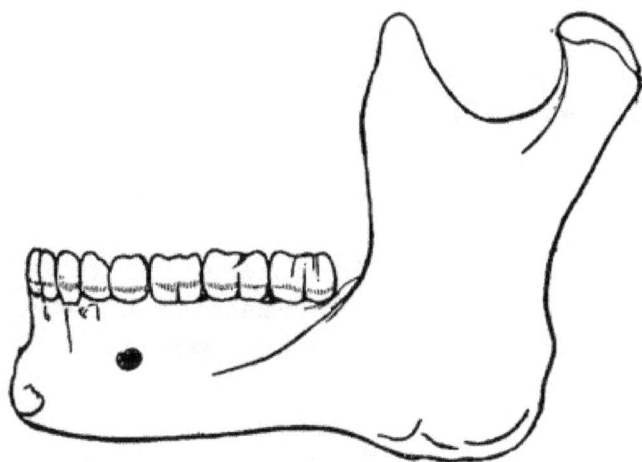

FIG. 85. — BLACKBOARD SKETCH. Side view of the Lower Jaw.

of blood, etc., — in other words, the difference between the
fresh (live) bone, and an old, dry one (dead) easily found for
the purpose.

2. *To show the Minute Structure of Bone.* — Show the
minute structure of bone by specimens of bone mounted on

glass slides, and sold for microscopic use, costing only a few cents. If a microscope cannot be had, let each pupil draw a blackboard-sketch of the microscopic look of bone.

3. *To show the Animal Part of Bone.* — Get a chicken's leg, or a sheep's rib. Scrape and clean. Put one of the bones to soak into a mixture of four tablespoonfuls of muriatic acid to one pint of water. A wide-mouthed bottle is the best thing to hold it ; next, an earthen bowl.

Soak from five days to a week. It can now be bent, twisted, and even tied into a knot, showing that the earthy matter has been dissolved.

4. *To show the Earthy, or Mineral, Part of Bone.* — Get a large soup-bone from the table. Roast it on a bright, hot coal-fire for three hours. Do it carefully, and get a good specimen free from bone-black. The animal matter has now been burned out. The earthy part, a white, brittle mass, is now seen, showing every outline of the bone. Crumble parts of it between the fingers, and otherwise experiment upon it.

*Study of the Skeleton.* — Let us now proceed to study the skeleton in two ways, — first, by means of the mounted skeleton prepared for such use ; second, by the examination of the skeleton or bones of our own person, or that of a friend.

Every important part of the body can be mapped out with the fingers. Whether a skeleton or a chart is used, this drill on the living body is essential to a practical knowledge of anatomy. Let a boy stand before the class ; and, as the teacher points out on his person the location of the various bones and bony landmarks of the body, each one of the class should stand, and do the same on his own person.

The amusement of the thing will soon wear away, then settle down to hard drill. Do it thoroughly. Crowd the

fingers deep down into the flesh, and trace the outline of the
bones (the collar-bone and shoulder-blade, for example) very
carefully. Do it all with neatness, tact, and a due regard for
the feelings of all concerned.

Use both the common and the scientific name for many of
the bones. Drill thoroughly until all become perfectly familiar with the location of the most important and accessible
bones and landmarks of the body.

5. *The Skeleton. — Its Study as a Whole.* — Point out head,
trunk, and limbs; long, short, flat, and round bones. Note
how some bones protect
certain parts, how others
are built for use, beauty,
and protection. Other
such points will readily
suggest themselves.

Do the same on the
person, *so far as it is convenient.* Use the chart for
the same purpose.

6. *Parts of the Skeleton.
— The Head.* — Drill on
the skeleton for the bones
of the head; the same on
the person. Use the chart.

7. Show how the bones
of the skull are jointed;

Fig. 86. — BLACKBOARD SKETCH. Upper
end of the Thigh-bone. Showing how the
girders of bone are arranged to sustain great
weight and render the bone less liable to
fracture.

the sutures; the plates of bones; the rough surfaces for the
attachment of muscles; holes and notches for arteries and
nerves. If nothing better can be had, get a part or the
whole of some animal's skull from the market, or even from
the fields.

8. *The Trunk.* — Note the various bones as before.

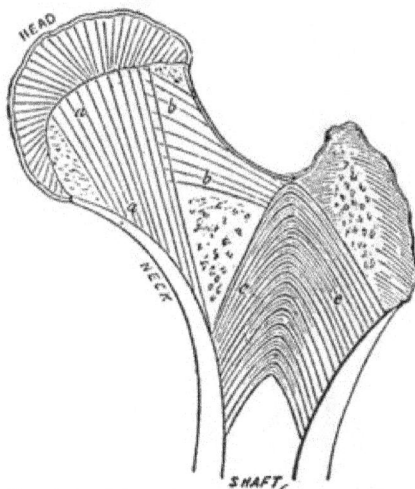

Trace and count each rib and vertebra. Make a special study of the spine.[1]

9. *The Upper Limbs.* — The upper arm, fore-arm, hand. Experiment exactly the same as before. Every part can be carefully traced on the person.

10. *The Lower Limbs.* — Thigh, lower leg, foot. Experiment as before.

11. *Repair of the Bones.* — Ask some pupil who has broken his arm or collar-bone some time previously, to show how well the broken parts have united; the healing-cement at place of union, etc.

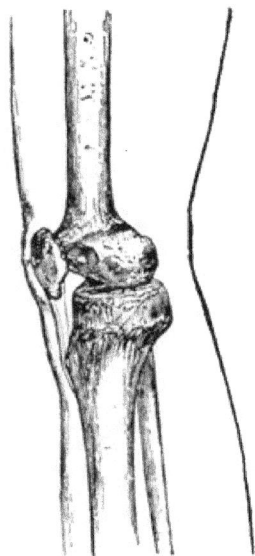

12. *Joints.* — Get a part of a calf's or sheep's leg at the market. Open up the joint carefully, and carry it fresh and moist into the class for experiment.

Note the sticky joint-oil; synovial fluid; glistening surfaces of the ends of the bones; the gristle; cartilages; place where tendons, ligaments, and muscles are fastened.

13. *To show the Various Kinds of Joints.* — Illustrate on the person the fixed and movable joints; ball and socket; hinge; pivot joints. Try the principal joints of the body, and see just how much and what motion they have.

FIG. 87. — BLACKBOARD SKETCH. The arrangement of bones at the Knee Joint.

14. *The Ligaments.* — In cutting open the joints, note the tough, firm, and gristly bands, which help hold the bones together. These are the ligaments. Try to tear or wrench them

---

[1] Some parts of this experiment, as well as others, should be reserved for the privacy of one's room. It is a most useful exercise to review the day's work after undressing for the night.

off from the bone, to see how firm and tough they are. Carefully dissect one or two ligaments in whole or in part, leaving one end fastened to the bone, to illustrate their action.

15. *Bony Landmarks of the Body.* — Go over the skeleton or person, and locate with the fingers certain important "bony landmarks," as the angle of the lower jaw, elbow-end of the ulna, wrist-end of the radius, highest point (acromion) of scapula, lower end of breast-bone, upper end of tibia; lower end of fibula, etc.

Other landmarks will suggest themselves. This will make an excellent review-exercise.

---

## EXPERIMENTS WITH THE MUSCLES

### [CHAPTER III. PAGE 42.]

EXPERIMENT 16. *To show the Gross Structure of Muscles.* — Get about one-half of a pound of lean corned beef, a strip with the fibres running all one way. Have it thoroughly boiled. Let it cool, and press it with the weight of several flat-irons. Put it on a firm board or table, and pick it in pieces with two darning-needles.

Note the connective tissue, the larger muscular fibres. Pick with the needles until the fibres are too small to manage. Continue with a hand magnifying-glass. Examine the tiniest bit of fibre with a microscope.

17. Again, boil a beef shin-bone for several hours. Note as before the coarse structure of the muscular fibres. Cut away the muscles and dissect, to examine the fibrous tissue, fat tissue, ligaments, and cartilage, or gristle.

18. *To show how Muscles contract and relax.* — Get the lower part of a sheep's leg at the market, with the foot or

hoof still on ; dissect with a sharp knife one or more muscles, leaving the insertion. Even if it is roughly done, it is no matter ; for the fibres can be carefully smoothed into place with a knife-blade, stained with carmine ink, and made to look natural as life.

Better make the dissection a week or so ahead of time, and let the parts harden a little in dilute alcohol. Then stain, and get it ready just at the time needed. The contraction and relaxation of muscles are thus roughly shown.

19. Use your own biceps and triceps muscles to show how muscles contract and relax, and how they oppose each other in action.

Repeat the same with other muscles, as the flexors and extensors of the fingers and toes, of the leg, etc.

20. *To show the Structure and Action of Tendons.* — Take some of the material from the previous dissection, and examine the tendons. Get the leg of a fowl at the market, to show the tendons which make the toes bend. Get the butcher to save the hoof of a calf or sheep with one end of the tendon of Achilles still attached. Dissect it, and test its strength.

21. Make a study of the tendons on your own person. Grasp the powerful tendons of the hip-muscles under the knee, of the lower leg near the ankle, of the fore-arm at the wrist.

Experiment with white braid tied on the tips of the fingers, and fastened on the arm, to show how the tendons of the arm work. Tie a ribbon tightly round the wrist to imitate the annular ligament, and note how differently the tendons act when they are under it or above it.

22. *To show how Muscles use the Bones as Levers.* — First use a blackboard-pointer, broom-handle, or any other familiar object as a lever. Practise using it as a lever of the three different kinds, until the principles are familiar.

Now illustrate on the person. Vary this experiment in many ways. Lift a book on the toes, by the fingers, on the back of the hand, by the mouth by bending over to reach it, raise the book from a chair, and rest it on the head.

Fasten a piece of braid to the neck of a bottle. Now tie one end to various parts of the body, as the middle of the arm, the fore-arm, the end of the fingers and foot, etc.; raising the bottle each time, and explaining exactly what kinds of levers are used, and also what kinds of muscles are used, — flexors or extensors.

23. *To locate some Important Muscles.* — Locate and describe them on the chart. Do the same as far as it is possible and convenient on the person.

Clutch tightly such muscles as the biceps, deltoid, great pectoral, muscles of the calf of the leg, etc. Run the fingers over the location of others.

Drill until the location and duty of the more important muscles become familiar.

24. *Muscular Landmarks.* — Use the chart, skeleton, and person to locate the most important muscular landmarks, as the origin and insertion of the biceps; tendon of Achilles; aunular ligament; edges of the great pectoral and biceps; flexor cords of the forefinger, etc.

This may be made an admirable drill-exercise.

---

## EXPERIMENTS ON FOOD AND DRINK

[CHAPTER V. PAGE 70.]

A series of most useful experiments may be made on the subject of food and drink. The common articles of diet greet us on every hand. It should be our object to understand the *principles* which underlie the matter of daily food.

We should become familiar with the principal substances contained in the three great classes of foods. We can do this by exhibiting specimens, and by experiment.

Specimens of the various cereals, starches, sugars, oil, etc., should be shown, which have been carefully collected, and kept for class-use in empty morphine or quinine bottles bought at the drug-store. Small radish or pickle bottles are good enough. Each specimen should be neatly labelled with its exact name.

Many interesting facts can be brought out about the practical use of these substances as foods, when we are able to touch, taste, smell, and see the substances themselves.

EXPERIMENT 25. *To show Albumen.* — The albumens are all rich in one or more of the following organic substances : albumen, casein, fibrine, gelatine, gluten, and legumen. Boil an egg hard. The white is *albumen* hardened by heat (see **Exper. 27**).

26. *To show Casein.* — Pour some liquid rennet, vinegar, or a little weak acid, into some milk. A whitish substance (the curd) separates from it. This nitrogenous substance is *casein*, the chief constituent of cheese.

27. *To show Fibrine.* — Take a piece of lean meat, and wash it thoroughly in water, squeezing and pressing it well in a lemon-squeezer. A whitish, stringy mass is obtained, which is the *fibrine.* The albumen is dissolved in the water. Boil the water after the meat has been washed. The heat coagulates the albumen.

28. *To show Gelatine.* — Boil a bone a long time. Most of the animal matter will be dissolved. The substance thus dissolved is *gelatine.*[1]

---

[1] You may have seen your mother boil calves' feet to make jelly for the sick-room. The familiar substances glue and size are simply gelatine obtained by boiling the hoofs, horns, skins, etc., of animals.

29. *To show Gluten.* — Put a handful of flour into a muslin bag, and squeeze it well in a basin of water.  The water becomes milky; while a sticky, yellowish-white substance remains in the bag.  This sticky substance is the *gluten.* Allow the water to stand, and the *starch* settles at the bottom in a white powder.

30. *To show Legumen.* — Boil a few peas or beans in the pod, until they become a sticky, pulpy mass.  This is the nitrogenous principle called *legumen*, and resembles the white of an egg and the gluten of flour.  Nearly every kind of vegetable is rich in this flesh-forming material.

31. *To show the Starches.* — Specimens of the more common kinds of foods rich in starch should be shown, and passed round the class in bottles.  Potato, rice, sago, arrowroot, tapioca, corn-flour, etc., may be used for illustration.

The presence of starch is easily proved.  Boil a small quantity of flour, rice, bread, potato, or arrowroot in a little water, in a test-tube.  Add a drop or two of the tincture of iodine, and the mixture will turn blue.

32. *To show the Sugars.* — *Cane-sugar* is familiar as cooking and table sugar.  Various specimens should be shown, including varieties of molasses, in two-ounce vials.  To show *grape-sugar*, get some raisins, and pick out a handful of the little white grains.  Get a cent's worth of *milk-sugar* at the drug-store.

In starch and sugar, the oxygen and hydrogen are combined in just the proportions necessary to form water.  Sulphuric acid reacts upon either starch or sugar to remove the oxygen and hydrogen; and carbon, or charcoal, is left behind.  Hence, the name *carbo-hydrates*, or hydrates of carbon, is given to this class of substances.  There are three groups, the *sugars proper*, the *glucoses*, and the *starches* and *gums*.  In a rough way, we may speak of them as *carbon dissolved in water*.

33. Take several wine-glasses.  Put some thin starch paste into the first, and several varieties of sugar into others.

Pour in carefully down the side of each glass a little sulphuric acid. The mixtures will turn black. The acid has
precipitated the charcoal. Stir a little charcoal in another
wineglass full of water, and pour in acid as before. It
remains black : no change has taken place.

34. By heat, starch is converted into a kind of sugar.
This is called *dextrine*. This is a gummy substance, soluble
in water, and is used on the back of postage-stamps.

Touch the tip of the tongue to the back of a postage-stamp,
and the sweet taste will be noticed. Dextrine is sold in the
shops under the name of " British gum."

35. *To show the Various Kinds of Fats and Oils.* — Show
specimens of fats that are solid at ordinary temperatures,
such as *beef-suet, mutton-suet, lard and butter.*

Liquid fats are commonly called *oils*. The two principal
kinds used as food are *olive-oil* and *cod-liver oil*. The first
is obtained from the fruit of a tree, and the second from the
livers of the cod and the haddock.

36. *Simple Experiments to show that Milk is a Model Food.*
— Milk is rightly called a compound food, since it is composed of at least five kinds of food-substances. Take some
milk "fresh from the cow," and place it in a tall, narrow
glass vessel, and allow it to stand for several hours. A
quantity of *cream* rises to the top. Cream is the milk-fat.
It is simply made up of tiny bags of fat, each of which has
a covering of curd.

By churning the cream about a short time, the covering is
broken, and the little lumps of fat unite to form the yellow
solid called *butter*. Now skim off the cream : what remains
is called *skim-milk*. Add a teaspoonful of vinegar, weak
acid, or liquid rennet, to the skim-milk. Solid whitish lumps
of *curd* will be seen separating from a watery fluid called
*whey* (see **Exper. 83**).

We have now divided the milk into three parts, — cream, curd, and whey. Cream is the fat ; curd is the casein, or nitrogenous part, which, when pressed and dried, is called *cheese ;* and whey is *milk-sugar* and *mineral matter* dissolved in *water.*

These simple experiments show that milk contains flesh-forming, bone-making, and heat-giving materials, and that it contains these in the right proportions, especially for children.

37. *To show Water, and the Minerals.* — *Water* is too familiar a food for experiment. If water could be easily obtained from some iron or sulphur springs, it would be interesting to compare the taste with specimens of hard and soft water from the neighborhood.

The mineral matter left from Experiment 4 shows the various kinds of *lime* and *potash.* An egg-shell is a familiar example of *carbonate of lime.* Various kinds of *salt* (forms of soda) can be obtained from any grocer.

The white of an egg contains a little *sulphur.* Leave a silver spoon in it for a short time, and the sulphur will blacken it.

---

## EXPERIMENTS WITH ALCOHOL

### [CHAPTER VI. PAGE 81.]

EXPERIMENT 38. *Alcohol, and how it looks.* — Get at the drug-store a four-ounce white-glass bottle, and let the druggist fill it for you with the best alcohol. Have it tightly corked, and properly marked with a label gummed on the bottom of the bottle.

To remember : Alcohol is a thin, colorless liquid, which looks like water.

39. *To show how Alcohol burns.* — Turn a little alcohol into an old-fashioned fluid lamp. Light it, and note the character of the light and the heat. Put a white saucer or plate closely down on the flame for a few minutes.

Note that the alcohol burns without soot, giving little light but great heat.

SUGGESTION. — If a fluid lamp is not easily obtained, use a small kerosene lamp without the chimney. A lamp good enough for simple experiments can be made out of a common mucilage bottle, using the hollow handle of the metallic brush for the wick. A piece of a bean-blower run through the cork of an empty horse-radish bottle, using a rolled piece of kerosene-wick for the wick, will also provide the necessary apparatus for an alcohol lamp.

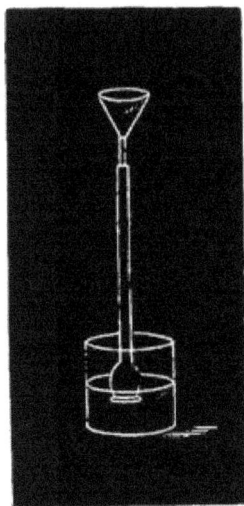

FIG. 88.

40. *To show that Alcohol and Water unite readily, with a Slight Decrease of Volume.* — Use a test-tube which will contain one fluid ounce (eight teaspoonfuls). Drop in thirty drops each of alcohol and water. Gum a strip of white paper four inches long and one-fourth of an inch wide to the side of the test-tube, bringing the lower end on a level with the sixty drops of fluid in the tube. Have the strip of paper marked off into inches and one-eighths of an inch.

Drop water carefully into the tube until exactly two inches have been added, as shown on the graduated paper ; then drop in alcohol until exactly two inches more have been added. Shake the tube carefully, taking care not to spill any of the fluid.

Note that the alcohol and water unite with a slight decrease of volume.

41. *To show the Great Attraction of Alcohol for Water.* — Tie tightly some thin membrane (like the coverings of the sausages sold in the markets) across the mouth of a bulb test-tube.   Place the test-tube (bulb downward) in a goblet or beaker.   Fill the bulb with alcohol, and pour pure water into the goblet or beaker (Fig. 95).

The alcohol will soon rise in the tube, thus proving that alcohol has such a liking for water that it has caused the water to filter through the membrane.

42. *To show the Origin of Alcohol in Fermented Liquors.* — Take a common glass fruit-jar which will hold one pint. Fill the jar one-half full of water, and add molasses until it is of a deep-brown color.   Add a teaspoonful of yeast or one-half of a yeast-cake.   Cover so as to admit some air, and keep at a temperature of 70° F. for a day or two.

Note the result: The mixture has the odor of alcohol. What has happened?   Why, the yeast (called a "ferment") has changed the sugar of the molasses to alcohol.   The process of changing is called "fermentation."

43. Take as before a common glass fruit-jar which will hold one pint.   Fill it one-half full of sweet cider.   Allow the liquid to stand exposed to warm air.   After a few days, shake the jar and it will be found that the liquid has the peculiar odor of alcohol.   Very minute ferments have changed the sugar of the sweet apple-juice to alcohol.   If the liquid should be left exposed to the air, it would, after a time, change to vinegar, which is known by its sour, acid taste.

44. Use a fruit jar as before.   Fill it one-half full of sweet or unfermented wine.   Continue the experiment as explained in Experiment No. 43.

45. Take two wide-mouthed bottles holding one pint each.   Be able to connect the two bottles *A* and *B*, as shown

in Fig. 96. Let a glass tube run from bottle *B*, and empty into a glass jar (*C*). Use a little putty or sealing-wax to stop any leaks made by the tubing running through the corks.

Fill the bottle *A* about one-half full of molasses. Add one yeast-cake which has been dissolved in water. Fill the bottle with water, and shake vigorously. Fill the bottle *B* nearly full of water.

Keep the apparatus in a warm room for two or more days: an even temperature of 75° to 80° F. will do.

The pupil will note that the fluid in the bottle *A* soon begins to "work," and that a substance called the lees set-

Fig. 89.

tles to the bottom of the bottle. The water in Bottle *B* will pass up the tube and over into the glass jar *C*. Note also that there is a change in the smell of the liquid. By the change that has taken place a gas has been formed which has forced the water in *B* over into the glass jar. This change is called fermentation.

46. *To show the Presence of Alcohol in Distilled Liquors.* — Heat hard cider or sherry wine in a large-sized test-tube over an alcohol lamp. Run a piece of glass tube, bent at right angles, through a cork which fits the test-tube : let the other

end of the tube be fitted into the cork of a wide-mouthed bottle which is set into a basin of cold water to condense the steam.

The resulting liquid has a marked odor and taste.  It is stronger than the fermented liquid.  This process is called "distillation," and the resulting liquid, "distilled liquor."

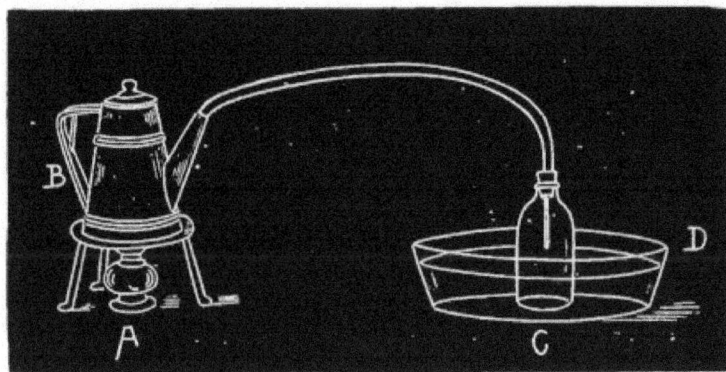

FIG. 90.

47. Pour one-half of a pint of hard cider into a common tin coffee-pot.   Fasten a piece of rubber tubing to the spout. Have the other end of the tubing run into a wide-mouthed bottle sunk into a basin of cold water.   Cloths wrung out in ice-water may be wrapped round the bottle.   Heat the cider in the coffee-pot by an alcohol lamp, placed as shown in Fig. 97.   Do not allow the cider to boil.

Note that the color and odor of the resulting liquor differ from the color and odor of the hard cider.   A stronger liquor has been produced by this process.   This process is called distillation.   The stronger liquor has been separated from the water.   It may be necessary to re-distil the resulting liquor several times before a liquor is found pure enough to burn.

48. *To show the Effect of Alcohol on Albumen.* — Place the white of an egg in an empty quinine or horse-radish bottle. Pour in some strong alcohol. Stir with a spoon or glass rod.

Note that the alcohol hardens, or coagulates, the albumen ; in other words, the alcohol has such a liking for the water of the albumen, that it withdraws it, leaving it hard.

49. Repeat the same experiment, using dilute alcohol (one-half alcohol and one-half water).

50. Place in the same kind of a bottle as before a small strip of raw beefsteak. Add strong alcohol, and let mixture be set aside for a few days.

Note that the meat seems hard. The alcohol has coagulated the albumen of the meat. The alcohol withdraws the water of the fresh, elastic, muscular tissue, leaving it tough and hard.

51. Squeeze a piece of fresh beef in a lemon-squeezer over a wide-mouthed bottle. Pour in a little water; stir it until well colored with blood. Add strong alcohol. Set aside for several days.

Note that the liquid is full of white particles. The alcohol has coagulated the albumen of the blood.

52. Repeat the last two experiments, using *dilute* alcohol. Note the result after a few days.

Take two bottles, into one of which some days before were placed weak alcohol and a piece of beef ; into the other, pure alcohol and beef. Take a piece of fresh beef, and you will find that it is as strong as a whip-cord. Now observe these pieces in the bottles. They are dry and brittle. They break like a piece of suet.

53. *To show the Action of Alcohol on Pepsin.* — Take two ordinary wide-mouthed bottles holding from a gill to half a pint (*A* and *B*). Put into the first bottle five grains of pepsin and two or three ounces of lukewarm water. Make it acid with twenty drops of strong muriatic acid.

Do exactly the same thing with the second bottle, except add three tablespoonfuls of alcohol. Set the bottles in any convenient place sufficiently warm (near a stove or on a mantel near a warm smoke-pipe) to keep the contents of the bottles at about blood-heat.

After ten hours, note the changes, if any. The pepsin is precipitated in the form of white, stringy particles in the bottle containing alcohol, while the other bottle shows no change; in other words, the alcohol has precipitated the pepsin of the artificial gastric juice.

54. Prepare the bottles exactly as before. To the bottle *A* add two teaspoonfuls of finely minced albumen. Put it in a warm place, as in Experiment No. 53. Shake very often. The pepsin will begin to act upon the albumen, gradually softening it. In actual digestion, albumen is thus made ready to soak through the moist lining of the stomach.

Do exactly the same thing with the second bottle (*B*) except add three tablespoonfuls of strong alcohol. Expose the bottle to heat, as before. After ten hours, note the result. It will be seen that alcohol has coagulated the albumen in bottle *B*.

55. Repeat Experiment No. 54. Instead of the white of an egg, add a teaspoonful of finely cut, cooked, and lean beefsteak or corned beef. In the bottle *A* the meat will be more or less dissolved or partially digested in twenty-four hours. In the bottle *B* the effect of the alcohol will be to make the meat seem hard. In short, the alcohol has coagulated the albumen of the meat.

56. Repeat Experiments 53 and 54. Instead of alcohol, use strong beer, whiskey, and rum, to show the action of the more dilute forms of alcohol upon the various types of albumen.

57. Repeat Experiment 41. Instead of plain water, add one-quarter of a teaspoonful of common salt to the given amount of water. This saline solution will represent, in a general way, the plasma of the blood. The alcohol will rise in the tube, as in Experiment 41. Now, if we use very dilute alcohol (one part of alcohol to ten parts of water), we find that the alcohol in the tube falls.

In other words, when alcohol is sufficiently dilute, it will go towards the fluid representing the plasma of the blood.

58. *To show how Alcohol coagulates the Blood.* — Get your market-man to carry a clean wide-mouthed fruit-jar to the slaughter-house, and let the butcher fill it with fresh blood. Add at once to the fresh blood a heaping tablespoonful of Epsom salts dissolved in a coffee-cup full of water. This strong saline solution will prevent the formation of a clot. Keep the whole mixture in a cool place, and do not shake it. Draw off two ounces of the mixture, and add to it a pint of water. Set it aside. It remains clear.

Dilute the same amount of the mixture with a pint of water, and pour in a few tablespoonfuls of strong alcohol. Note that the blood is soon coagulated.

59. *To show the* **Effect** *of Alcohol* **on the** *Blood Corpuscles.* — Squeeze the end of the forefinger and pierce the skin quickly with a fine needle, drawing a drop of blood. Examine it carefully with a microscope. Note the little corpuscles, their shape, and how they are arranged.

Add the tiniest drop of alcohol. The corpuscles shrink and become of a different or irregular shape. The alcohol has coagulated the albumen of the corpuscles.

60. *To show the Effect of Alcohol on* **the Circulation of the** *Blood.* — Stretch the web of a frog's foot over a hole in a · thin board, as fully directed in Experiment No. 98. Do this carefully to avoid giving pain to the frog.

Note carefully the circulation of blood through the very small blood-vessels.

Now put on the web a drop of dilute alcohol, and note the result. The little blood-vessels seem to stretch and to allow the blood corpuscles to pass through more rapidly.

The alcohol has weakened the nerves which regulate the flow of blood in the capillaries, thus allowing more blood to flow through them.

61. Repeat the preceding experiment. Instead of dilute use strong alcohol. Note that the capillaries soon shrivel, and the flow of blood stops.

The alcohol in this case has paralyzed the nerves which control the capillaries. The tiny blood-vessels now contract, and stop the flow of blood.

---

## EXPERIMENTS ON DIGESTION

### [CHAPTER VII. PAGE 92.]

62. *To show the Anatomy of Certain Organs, and Parts of Organs, of Digestion.* — Have each pupil examine his own mouth, and also that of some friend. Note the lips, tongue, hard palate, soft palate, uvula, tonsils, and upper part of the gullet.

63. *The Teeth.* — Get a specimen of each kind of tooth if possible. A dentist-friend will give you what you need. Use a very fine saw to saw a perfect molar in two lengthwise. If need be crack the tooth with a hammer.

Note its structure in a general way, — its crown, body, cusps, roots, enamel, dentine, pulp-cavity, etc.

64. Make a blackboard-sketch of a tooth on a large scale, using colored crayon to make plain the various parts of a tooth.

65. *To show the Difference of Teeth in Various Animals.* — Get, if possible, specimens of teeth of the more familiar animals,[1] as the cat, dog, rat, squirrel, etc. Compare them with the specimens of human teeth.

66. With the help of a mirror, let each pupil locate his own teeth. Note the incisors, eye-teeth, bicuspids, molars, and "wisdom" teeth if any. In the same way, note the teeth of some schoolmate.

67. Locate as near as possible the position of each salivary gland. Press on the part of the cheek opposite the second small molar of the upper jaw, and notice the increased flow of saliva. The appearance of saliva is familiar.

68. *To show Location of Other Organs of Digestion.* — Make a blackboard diagram, in the rough, of the organs of digestion, beginning at the cardiac end of the stomach, after having studied the digestive organs on the chart.

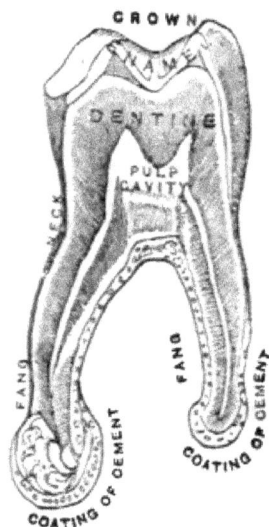

FIG. 91. — BLACKBOARD SKETCH. Section of a Tooth.

Map out on the person in a very general way the location of the stomach, small intestines, and the colon, liver, pancreas, and spleen. The location of the stomach and liver is especially important.

69. *To show how the Wall of the Stomach looks.* — The wall of the pig's stomach resembles the human stomach. Get from the market a piece of a pig's stomach. Cut off bits of it, and examine it thoroughly. Scrape off the inner or mucous coat with the edge of a very sharp knife.

---

[1] No stress is laid in this book upon the subject of "comparative anatomy." Many valuable points can be given orally by the painstaking teacher. For reference, consult MIVART'S *Lessons on Elementary Anatomy.*

Use a magnifying-glass; find the openings of the gastric tubes. Pick with fine needles, and note the muscular coats. Contrast the pig's stomach with that of a cow's by examining a piece of tripe.

70. *To make the Saliva flow.* — Think of some favorite article of food (both at the time you are hungry and when you are not), and note the flow of saliva. Push a lead-pencil or the finger to and fro in the mouth several times, and note the flow of saliva.

Chew an oyster-cracker, and note carefully how it is moistened with saliva. Grind up several crackers rapidly in the mouth. Note how difficult it is to swallow, or even chew, when there is little saliva.

71. The importance of the saliva in favoring the movements of chewing may be illustrated by the experiment of wiping the inside of the mouth perfectly dry with a towel or handkerchief; when it will be found almost impossible to move the jaws until the saliva is again secreted in sufficient quantity to moisten the surface of the tongue, cheeks, and gums.

72. The importance of the saliva as a solvent may be shown by wiping the upper surface of the tongue quite dry with a napkin, and then placing a small quantity of powdered sugar upon it. The sugar will be found to be as destitute of taste as so much sand; whereas in its ordinary moist condition the tongue perceives the taste of sweet substances very distinctly.

This experiment also illustrates the fact that the terminations of the nerves of taste can be affected only by substances brought in contact with them in the liquid state.

73. *To show the Action of Saliva on Starch.* — Chew slowly a piece of fresh bread.[1] Note how sweet it tastes after

---

[1] Chew pieces of the brown crust of the bread. It is quite sweet, and readily dissolves; because, exposed to more heat than the rest of the loaf, the starch has been changed into dextrine before the bread left the oven. Hence, crust and toast are favorite articles of food, especially with old people.

it is well wet with the saliva.  Do the same with a mouthful of paste made of pure arrowroot (almost pure starch), made with boiling water, and allowed to cool.

74. Take three test-tubes.  Into test-tube No. 1, half full of cold water, add some half a dozen drops of the arrowroot paste; now put one of Fehling's Test-Tablets.[1]  Mix thoroughly, and boil over a spirit-lamp.  There is no turbid brick-red deposit to show that there is sugar in the starch (arrowroot).

75. Put several teaspoonfuls of saliva into test-tube No. 2.  Dilute it with water.  Add a test-tablet, mix thoroughly, and boil.  There will be a violet color, but no brick-red deposit to show the presence of sugar in the saliva.

76. Take test-tube No. 3, and put a few drops of the starch paste into a teaspoonful of saliva, and fill the test-tube half-full of water.  Put the mixture away for ten minutes in a warm place.  Now add a test-tablet.  Mix and boil.  A turbid brick-red deposit will be thrown down, showing that sugar is present, due to the action of the saliva on starch.

77. *To show the Action of Gastric Juice on Albumen.* — Take an ordinary four-ounce bottle with as wide a mouth as convenient.  Put into it the following: Pepsin in scales (Fairchild's),[2] one grain, and four tablespoonfuls of lukewarm water.  Make it acid with ten drops of strong muriatic

---

[1] These tablets have been used of late as a convenient means by which physicians may test for sugar.  It is simply an easy way to use the test usually made with a solution of caustic potash and blue vitriol.

[2] It is not at all necessary to go through the long **process of** getting pepsin from a pig's stomach.  Pure pepsin, obtained from the stomach of the pig, is now made by dealers for medical use; and the necessary amount **can** be bought of a good druggist for a few **cents.**  Use only *pure* pepsin in **the** experiments.  **The** student is advised to use Fairchild's " Pepsin " and " Extract **of Pancreas.**"

acid. Add two teaspoonfuls of finely minced egg-albumen.[1]

Set the mixture in any convenient place sufficiently warm, as in a basin of warm water, near the stove, or on the mantel near the warm smoke-pipe, to keep the contents of the bottle at about "blood-heat."[2] Shake every few minutes. As soon as the whole mixture becomes of a proper temperature, the pepsin will begin to act upon the albumen; and it may be seen to be gradually softened and digested.

This dissolved mass is now in a condition to readily soak through the moist lining of the stomach, and is known as "peptone."

78. Repeat the same experiment, using half a teaspoonful of finely cut, cooked, and lean beefsteak or corned beef instead of the white of an egg. The meat will be more or less dissolved, or partially digested, in twenty-four hours.

This is only a crude way to show the action of gastric juice on meat-albumen.

79. Get some freshly drawn blood. Whip it with twigs. Pull off the collected fibrine, and wash it carefully in clean water. Save this in dilute alcohol, and use it as a handy albumen for digestive experiments. Repeat as directed in Experiments 77 and 78.

80. *To show the Action of Bile on Fat.* — Shake up a little sweet-oil and water. "Oil and water will not mix," as the saying goes. Get the butcher to bring you a small

---

[1] The albumen or the white of an egg is taken as the *type* of albumen in a naturally pure state, and is well adapted to illustrate the peptonizing action of the gastric ferment. It should be first made ready as follows: Separate the white of a *hard-boiled* egg, and rub it through a coarse sieve, so as to divide it into small particles.

[2] In these experiments on digestion, the *temperature* should never be *below* blood-heat, and never allowed *above* a point at which fluids can be *easily borne by the mouth.*

bottle of bile (ox-gall), or, better still, ask him to bring the gall-bladder itself. Cut it open, and bottle the contents. Shake up some oil with bile, and a creamy mixture called an "emulsion" results.

81. *To show the Action of Pancreatic Juice upon Oils or Fats.* — Put two grains of Fairchild's Extract of Pancreas into a four-ounce bottle. Add half a tablespoonful of warm water, and shake well for a few minutes; then add a table spoonful of cod-liver oil; shake vigorously.

A creamy, opaque mixture of the oil and water (called an "emulsion") will result. This will gradually separate upon standing, the pancreatic extract settling in the layer of water at the bottom. It will again form an emulsion when shaken.

82. *To show the Action of Pancreatic Juice on Starch.* — Make a *smooth* starch paste, just as the laundress does in starching her clothes. Put two tablespoonfuls of this paste into a goblet or tumbler, and, while still so *warm* as to just be borne by the mouth, stir into it two grains of the Extract of Pancreas. The starch paste will at once become thinner, and gradually changed into soluble starch, in a perfectly fluid solution.

83. *To show the Action of Pancreatic Juice upon the Albuminous Ingredients* (casein) *of Milk.* — Into a four-ounce bottle put two tablespoonfuls of cold water; add one grain of Fairchild's Extract of Pancreas, and as much baking-soda as can be taken up on the point of a penknife. Shake well, and add four tablespoonfuls of cold, fresh milk. Shake again.

Now set the bottle into a basin of hot water (as hot as can be borne by the hand), and let it stand for about forty-five minutes. While the milk is digesting, take a small quantity of milk in a goblet, and stir in ten drops or more of vinegar. A thick curd of casein will be seen (see **Exper. 36**).

Upon applying the same test to the digested milk, no curd will be made. This is because the pancreatic ferment (called trypsin)[1] has digested the casein into " peptone," which does not curd. This digested milk is therefore called " peptonized milk."[2]

84. *Study of the Liver.* — The whole of a liver is rather an awkward and bulky thing to handle ; but it is a good plan to get a sheep's liver, and devote a little time to its gross anatomy. Take pains to get the liver with the gall-bladder unbroken. Having once secured a fine specimen, it may be kept for future use in dilute alcohol.

85. *To illustrate the Principle of Absorption.* — The manner in which absorbent vessels suck in juices may be illustrated in a general way as follows : Take a glass tube, and tie a piece of bladder, in the form of a small bag, at the end of it ; pour some mucilage into it, and then plunge the end into a cup or glass containing water.

` In a few moments the mucilage in the tube will begin to rise, and the water in the glass will sink ; because, being more dense than the water, it draws it in through the bladder, and will continue to do so until they are equal. Again, if we put mucilage into the glass, and fill the tube with water, the water will fall in the tube until they become of the same thickness.

It is on this principle that the absorbent vessels suck in liquids which are thinner than what they previously contained.

[1] The active principle of the saliva which acts upon starch is often spoken of as *ptyalin*, in order to distinguish it from the starch-digesting principle of the pancreatic juice. Both have the same action upon starch. Experiment 57 will also illustrate the action of saliva.

[2] Infants, especially those brought up on artificial foods, often suffer from ill digestion, "summer complaint," and other intestinal troubles during hot weather. Partially digested, or peptonized milk is often found to be of the greatest service at such times.

## EXPERIMENTS ON THE HEART AND THE CIRCULATION

[CHAPTER VIII. PAGE 122.]

EXPERIMENT 86. — Tie a string tightly round the finger, and pierce the tip of it with a needle. The blood runs freely, is red and opaque. Put a drop of fresh blood on a sheet of clean white paper, and note that it looks yellowish.

87. Put a drop of fresh blood on a clean white plate. Cover it with an inverted goblet or sauce-plate. Take off the cover in five minutes, and the drop has set into a jelly-like mass. Breathe into the goblet several times to moisten it, and replace it over the blood. Take it off in half an hour, and a little clot will be seen in the watery serum.

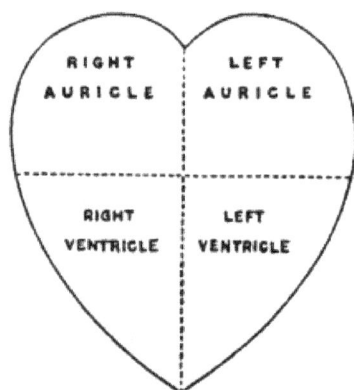

FIG. 92. — BLACKBOARD SKETCH. The Heart and its four rooms.

88. *To show the Blood-Clot.* — Carry a clean wide-mouthed bottle to the slaughter-house. A half-pound chloral bottle bought at any drug-store is the best. A clean horse-radish bottle will do. Fill it with fresh blood. Carry it home with great care, and let it stand over night. The next day the clot will be seen floating in the nearly colorless serum.

89. *To show the Fibrine of the Blood.* — Get a wide-necked bottle full of fresh blood. A pint or quart fruit-jar is the best for this purpose. Beat the blood vigorously for five minutes with some dry twigs, and hang them up to dry. Let the blood stand until the next day. Note that it has

not clotted. Wash out the twigs thoroughly, and a stringy mass of elastic substance will be found hanging to them. This is the fibrine.

90. Take some of the serum saved from Experiment 88, and note that it does not coagulate of itself. Boil a little in a test-tube over a spirit-lamp, and the albumen will coagulate.

91. To show that blood is really a mass of red bodies, floating in a fluid, which give it the red color. Fill a clean white-glass bottle two-thirds full of little red beads, and then fill the bottle full of water. At a short distance the bottle appears to be filled with a uniformly red liquid.

92. *To show the Blood Corpuscles.* — A moderately powerful microscope is necessary to examine blood corpuscles. Let a small drop of blood (which may be readily obtained by pricking the finger with a sharp needle) be placed upon a clean slip of glass, and covered with a piece of thin glass, such as is ordinarily used for microscopic purposes.

The blood is thus spread out into a film, and may be examined with the microscope. At first the corpuscles will be seen as pale, disk-like bodies floating in the clear fluid. After a time they will be observed to stick to each other by their flattened faces, so as to form rows. The colorless corpuscles are to be seen among the red ones, but they are much less numerous.

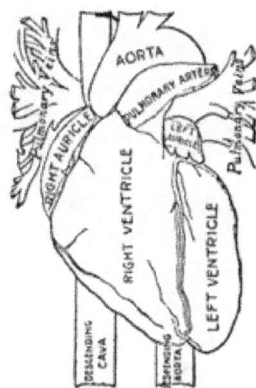

FIG. 93. — BLACKBOARD SKETCH. The Heart and its great Blood-vessels.

93. *To show how Blood holds a Mineral Substance in Solution.* — Put an egg-shell crushed fine into a glass of water made acid by a teaspoonful of muriatic acid. After an hour or so the egg-shell will disappear, having been dissolved in

the acid water. In like manner the blood holds various minerals in solution.

94. *To show the Sounds of the Heart.* — Locate the heart exactly. Note its beat. Borrow a stethoscope from some physician. Listen to the heart-beat of some friend. Note the sounds of your own heart in the same way.

95. *To find the Pulse.* — With your right hand grasp the right wrist of a friend, pressing with three fingers over the radius. Press three fingers of your right hand over the radius in the left wrist, to feel your own pulse. With the aid of a watch, count the rate of your own pulse per minute. Do the same with a friend's pulse.

96. *To show how the Pulse may be studied by pressing a Mirror over the Radial Artery.* — Press a small piece of looking-glass upon the wrist over the radial artery in such a way that with each pulse-beat the mirror may be slightly tilted. If the wrist be now held in such a position that sunlight will fall upon the mirror, a spot of light will be reflected on the opposite side of the room, and by its motion upon the wall will be shown the movements of the artery as the pulse-wave passes through it. — BOWDITCH'S *"Hints for Teachers of Physiology."*

FIG. 94. — How the Pulse may be studied by pressing a Mirror over the Radial Artery.

97. *Dissection of a Sheep's Heart.* — Get a sheep's heart with the lungs attached, as the position of the heart will be better understood. Let the lungs be laid upon a dish so

that the heart is uppermost, with its apex turned towards the observer.

The line of fat which extends from the upper and left side of the heart downwards and across towards the right side, indicates the division between the right and left ventricles.

Examine the large vessels, and, by reference to the text and illustrations, make quite certain which are the *aorta*, the *pulmonary artery*, the *superior* and *inferior venæ cavæ*, and the *pulmonary veins*.

Tie variously colored yarns to the vessels, so that they may be distinguished when separated from the surrounding parts.

Having separated the heart from the lungs, cut out a portion of the wall of the *right ventricle* towards its lower part, so as to lay the cavity open. Gradually enlarge the opening until the *chordæ tendineæ* and the flaps of the *tricuspid valve* are seen. Continue to lay open the ventricle towards the pulmonary artery until the *semilunar valves* come into view.

The pulmonary artery may now be opened from above so as to display the upper surfaces of the semilunar valves. Remove part of the wall of the right auricle, and examine the right auriculo-ventricular opening.

The heart may now be turned over, and the *left ventricle* laid open in a similar manner. Notice that the mitral valve has only two flaps. The form of the valves is better seen if they are placed under water, and allowed to float out. Observe that the walls of the *left* ventricle are much thicker than those of the *right*.  -

Open the left auricle, and notice the entrance of the *pulmonary veins*, and the passage into the ventricle.

The ventricular cavity should now be opened up as far as the aorta, and the semilunar valves examined. Cut open the aorta, and notice the form of the *semilunar valves*.

98.  *To show the Circulation in a Frog's Foot.* — In order
to see the blood circulating in the membrane of a frog's foot
it is necessary to firmly hold the frog.    For this purpose
obtain a piece of soft wood, about six inches long and three
wide, and half an inch thick.    At about two inches from one
end of this, cut a hole three-quarters of an inch in diameter,
and cover it with a piece of glass, which should be let into
the wood, so as to be level with the surface.    Then tie up
the frog in a wet cloth, leaving one of the hind-legs outside.
Next, fasten a piece of cotton to each of the two longest
toes, but not too tightly, or the circulation will be stopped
and you may hurt the frog.

Tie the frog upon the board in such a way that the foot
will just come over the glass in the aperture.    Pull carefully
the pieces of cotton tied to the toes, so as to spread out the
membrane between them over the glass.    Fasten the threads
by drawing them into notches cut in the sides of the board.
The board should now be fixed by elastic bands, or by any
other convenient means, upon the stage of the microscope,
so as to bring the membrane of the foot under the object-
glass.

The flow of blood thus shown is indeed a wonderful sight,
and never to be forgotten.   The membrane should be occasionally moistened with water.

Care should be taken not to occasion any pain to the frog.

---

## EXPERIMENTS ON THE LUNGS AND BREATHING.

### [CHAPTER IX. PAGE 142.]

EXPERIMENT 99.    *To locate the Lungs.* — Mark out the
boundaries of the lungs by "sounding" them; that is, by
percussion as it is called.   It simply means, to put one finger across the chest or back, and to give it a quick, sharp

rap with another. Note where it sounds hollow; that is, is resonant. This experiment can be done only in a very crude way.

100. Borrow a stethoscope, and listen to the respiration directly below the collar-bones. Note the difference in inspiration and expiration.

101. Get a sheep's lungs, with the windpipe attached, from a butcher or marketman. Ask for the heart and lungs all in one mass. Take pains to examine the specimen first, and pay only for a good one. Parts are apt to be hastily snipped or mangled.

102. Examine the windpipe. Note the horseshoe-shaped rings of gristle in front, which serve to keep it open.

103. Examine one bronchus, carefully dissecting away the lung-tissue with curved scissors. Follow along until small branches of the bronchial tubes are reached. Take time for the dissection, and save the specimen in dilute alcohol. Put pieces of the lung-tissue in a basin of water, and see if they float.

104. *To show how the Lung may be filled with Air.* — Take the other lung, and tie a glass tube six inches long into its bronchus. Attach a piece of rubber to one end of the glass tube. Now blow up the lung, and then let it collapse several times. When distended, examine every part of it.

105. *To show the Diaphragm.* — Dissect a rat or kitten, and cut out a section of the body large enough to show the diaphragm and its attachments. The parts can be neatly held in place by pins and silk threads, and the whole kept in dilute alcohol.

106. *To show the Vocal Cords.* — Get a pig's windpipe in perfect order from the butcher, to show the vocal cords. Once secured, it can be kept for an indefinite time in alcohol.

107. *To show that the Air we breathe out is warm.* — Breathe on a thermometer for a few minutes. The mercury will rise rapidly.

108. *To show that it is moist.* — Breathe on a mirror, or a knife-blade, or any polished metallic surface.

109. *To show that the Air we breathe out contains Carbonic Acid.* — Put a glass tube into a bottle of lime-water, and breathe out into it through the tube. The liquid will soon become cloudy, because the carbonic acid of the expired air throws down the lime held in solution.

110. *To show the Natural Temperature of the Body.* — Borrow a physician's thermometer, and take your own temperature, and that of several friends, by putting it under the tongue, closing the mouth tightly, and holding it there for five minutes.

111. "A substitute for a clinical thermometer may be readily contrived by taking an ordinary house thermometer from its tin case, and cutting off the lower part of the scale so that the bulb may project freely. With an instrument thus arranged, the pupils may take their own and each other's temperature; and it will be found that whatever may be the season of the year or the temperature of the room, the thermometer in the mouth will record about 99° F. Care must, of course, be taken to keep the thermometer in the mouth till it ceases to rise, and to read it while it is still in position." — Prof. H. P. Bowditch.

---

### EXPERIMENTS ON THE SKIN

[Chapter X. Page 161.]

Experiment 112. *To show the Gross Structure of the Skin.* — Examine your own skin in a general way. Stretch and pull it, to show how elastic it is. Get some idea of the scarf-skin by scraping off some with a sharp knife.

Note any white scars, liver-spots, etc., on the skin. Examine the skin with much care, with the aid of a good magnifying-glass. Note the papillæ on the palm of the hand.

113. *Hair and Nails.* — Pull a stout hair from the head, and examine it carefully with the magnifying-glass. Make a similar study of the finger-nails.

114. *The Sweat-glands.* — Use a magnifying-glass to study the openings of the sweat-glands, especially on the palms of the hands.

---

## EXPERIMENTS ON THE NERVOUS SYSTEM

### [CHAPTER XI. PAGE 177.]

EXPERIMENT 115. *To show the Brain and Spinal Cord.* — It is a troublesome matter to dissect out the brain and the spinal cord of one of the smaller animals like the rat, rabbit, or cat. The back of the animal to be dissected should be turned upwards, and the limbs should be securely fastened to the top of a small table.

The skin is removed from the head by making a longitudinal incision in it, and then turning it back upon each side. The muscles upon the sides and back of the skull should next be cleared away. A hole may now be made in the top of the skull with the point of a pair of scissors, and the bone broken away piece by piece, with a pointed pair of nippers, until the upper surface and sides of the brain can be seen.

In the cat, kitten, or rabbit, the cerebral hemispheres do not, as in man, cover the cerebellum ; and consequently the latter is seen when the brain is viewed from above.

When the bone has been broken away enough to expose

the sides of the brain, the latter should be carefully raised by inserting the tips of the fingers under it on one side, and gently pulling.

Several white cords will thus be seen passing from the under surface of the brain to the floor of the skull: these are the *cranial nerves.*

To expose the *spinal cord,* the skin of the back must be removed, and the muscles cut away from each side of the backbone, as close to the bone as possible; then, beginning at the back of the skull, the spinal canal is to be laid open by cutting with the nippers the two sides of each vertebra.

Care must be taken not to cut or tear the spinal cord, or the nerves passing off from it.

When the spinal cord has thus been exposed, and the membranes which surround it made out, the *spinal nerves* should be examined.

The brain and spinal cord should now be removed, by carefully lifting them, and cutting off the nerves by which they are attached.

The specimen should now be neatly fastened to a thin strip of smooth pine-wood, and firmly held in place with pins. The whole can be kept in alcohol and water. Great patience is necessary to make the dissection a success.

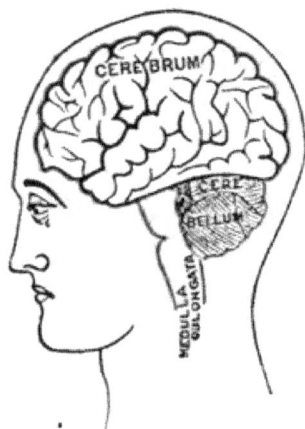

FIG. 95. — BLACKBOARD SKETCH. Showing the Position of the Brain.

116. *To show the Brain.* — A sheep's or calf's brain is to be preferred, on account of its larger size. Get one fresh from the butcher. Pay him to dissect away the skin and muscles of the skull, under your direction, and to saw open the cranium in a circular direction. Take time, and remove

the sawed top with great care, tearing away the dura mater from the bones.

Now cut away enough of this membrane so that the sides of the skull can be sawed and torn away, to allow us to lift out the brain, with proper dissection, in as perfect a state as need be. Put all the torn parts and membranes back into place. Put the whole aside to harden for three days in alcohol. Now we are ready to study its gross anatomy.

Note the *dura mater*, the *arachnoid* lining it, and the *pia mater* closely attached to the brain. Find the *cerebrum*, or big brain, the *cerebellum*, or little brain, the *medulla*, and the stumps of the *cranial nerves*, especially the stumps of the optic nerves. The hardened brain should be first examined as a whole, and compared with the description given in the text, or the pictures of the human brain.

117. *Sensations are referred to the Ends of the Nerves.* — Strike the elbow-end of the ulna against anything hard ("hitting the crazy-bone") where the ulna nerve is exposed, and the little finger and the ring-finger will tingle, and become numb.

118. *Every Nerve is independent of any Other.* — Press two fingers closely together. Let the point of the finest needle be carried ever so lightly across from one finger to another, and we can easily tell just when the needle leaves one finger, and touches the other.

119. *One Part of the Body works for the Good of another Part.* — Tickle the inside of the nose with a feather. This does not interfere with the muscles of breathing; but they come to the help of the irritated part, and provoke sneezing to clear and protect the nose.

120. *To paralyze a Nerve temporarily.* — Throw one arm over the sharp edge of a chair-back, bringing the inner edge of the biceps directly over the edge of the chair. Press

deep and hard for a few minutes. The deep pressure on the nerve of the arm will put the arm "asleep," causing numbness and tingling. The leg and foot often "get asleep" by deep pressure on the nerves of the thigh.

---

## EXPERIMENTS ON THE SPECIAL SENSES

### [CHAPTER XII. PAGE 210.]

EXPERIMENT 121. — Cross the middle finger over the forefinger. and shut the eyes. Hold a marble between the tips of the crossed fingers. This experiment will produce the impression of there being two marbles.

122. Shut both eyes, and let a friend run the tips of the fingers first lightly over a hard, plane surface, then pressing hard, and then lightly again, and the surface will seem to be hollowed out.

These two experiments show that we are easily misled, even by the sense of touch. It is mainly matter of habit and education.

123. Put a drop of vinegar on a friend's tongue, or your own. Notice how the papillæ of the tongue start up.

124. Rub different parts of the tongue with the pointed end of a piece of salt or gum-aloes, showing that the *back* of the tongue is most sensitive to salt and bitter substances.

125. Repeat the same with some sweet or sour substances, to show that the *edges* of the tongue are the most sensitive to these things.

126. Apply the points of a pair of compasses to various places on the surface of the body, to show the most sensitive spots.

127. We often fail to distinguish between the sense of

taste and that of smell. Chew some pure, roasted coffee, and it seems to have a distinct taste. Pinch the nose hard, and there is little taste. Coffee has a powerful odor, but only a feeble taste. The same is true of garlic, onions, and various spices.

128. Light helps the sense of taste. Shut the eyes, and palatable foods taste insipid. Pinch the nose, close the eyes, and see how palatable a teaspoonful of cod-liver oil becomes.

129. Gently turn the inner part of the lower eyelid down-

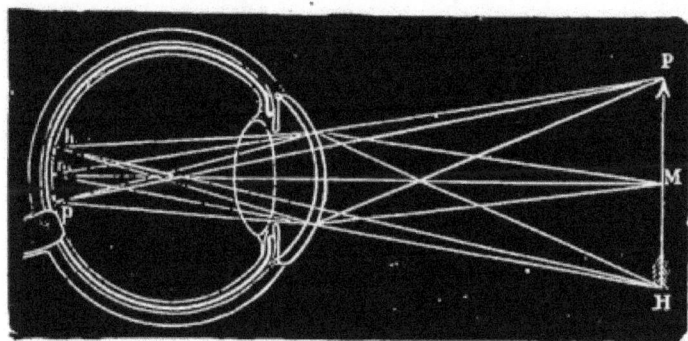

FIG. 96. — BLACKBOARD SKETCH. Showing how the Image of an Object is brought to a Focus on the Retina.

wards. Look in a mirror; and the small "lachrymal point," or opening into the nasal duct, may be seen in your own eye.

130. Close the eyelids, and press firmly on to the eyeball with two fingers. Note how firm and dense it is to what one might imagine.

131. *The Retina is easily tired.* — With a hand-mirror reflect the sunlight on the white wall, and keep it fixed. Look steadily at it for a full minute, and then let it suddenly be removed. Its "complementary" color — a dark spot — will appear.

Pin a round piece of bright red paper (large as a dinner-

plate) to a white wall by a single pin. Fasten a long piece
of thread to it, so it can be pulled down in a second. Gaze
steadily at the red paper. Have it removed while looking
at it intently, and a greenish spot takes its place.

132. *To illustrate the Action of the* **Crystalline Lens.** —
Hold a burning-glass in front of an open window in such a
way that the image of some object outside will be brought
to a focus upon a piece of paper. This miniature image
will be upside down.

133. The walls of the eyeball of a white rabbit are quite
transparent. Hold a lighted candle before the cornea. An
image of the candle will be seen reflected on the retina.

134. "Remove the front convex lens from a pair of opera-
glasses, or procure a convex lens with a gradual curve.
Hold it opposite a window, and place a piece of white paper
behind it to act as a screen. A small reversed picture of
the window-frame will appear on the paper.

"If the paper be moved to a certain distance, varying with
each lens, the picture will become clear and distinct, yet
with color-rings about the edges. At that distance from the
lens the paper is said to be *in focus*. If the paper be moved
nearer to, or farther from, the lens, the picture becomes
blurred, and the paper is said to be *out of focus*.

135. "Visit a photographer's studio. Request him to
point out and name the uses of the essential parts of the
camera, — the blackened box, the ground-glass screen, the
lens, the diaphragm, and the apparatus for adjusting the lens
and the screen to the object. Watch him place the camera
and then work the ground-glass screen into the proper focus.
When all is ready, put your head under the curtain of the
camera and study the reversed image depicted on the glass."
— CUTTER'S *Physiology.*

136. To vibrate the drum-membrane and the little ear-

bones, shut the mouth, and pinch the nose tightly. Try to force air through the nose. The air dilates the eustachian tube, and air is forced into the ear-drum. The distinct crackle, or clicking sound, is due to the movement of the ear-bones and membranes.

137. *The Retina is not sensitive where the Optic Nerve enters the Eyeball.* — This is called the "blind spot." Put two ink-bottles about two feet apart, on a table covered over with white paper. Close the left eye, and fix the right steadily on the left-hand ink-stand, gradually varying the distance of the eye to the ink-bottle. At a certain distance the right-hand bottle will disappear; but nearer or farther than that, it will be plainly seen.

138. Show the same thing in this way: make two black spots, with pen and ink, as large as two peas, upon a white

FIG. 97. — Diagram to show the Existence of the Blind Spot.

card, about three inches apart, varying the distance as before. At a distance of about six inches the "blind spot" will be noticed.

139. *Impressions made upon the Retina do not disappear at once.* — Look steadily at a bright light for a moment or two, and then turn away suddenly, or shut the eyes. A gleam of light will be seen for a second or two.

140. Take a round piece of white card-board the size of a saucer, and paint it in alternate rings of red and yellow, — two primary colors. Put it on the sewing-machine wheel, or glue it to a grindstone, and rotate it rapidly. The eye perceives neither color, but orange, — the secondary color.

141. To note the shadows cast upon the retina by opaque

matters in the vitreous humor, popularly known as floating specks, or gossamer threads, look through a small pin-hole in a card at a bright light covered by a ground-glass shade.

142. Test the eyes of the pupils by a set of *Snellen's test-types*. The following test-line should be easily read at ten feet. Some pupils may be able to read it at a distance of eleven or more feet.

# V Z B D F H K O S

143. *To dissect an Eye.* — The eyes of a codfish, a sheep, or a pig are easily obtained. Put them in the snow over night during winter, or freeze them with salt and ice during warm weather. They are easier to dissect when frozen.

Examine one specimen from the outside, however, before freezing dulls its bright colors. Saw out a square piece of the skull that holds the eye, and see just how it is held in place naturally, how the optic nerve enters the orbit, etc.

Dissect with great care the frozen eye.

The fat should be dissected off, together with the muscles, and the eye cut in half from above downward. This will be most easily accomplished by holding the eye between the forefinger and thumb of the left hand, upon a slice of turnip or potato, and then cutting it between finger and thumb with a sharp razor.

The transparency of the cornea, lens, vitreous humor, and also of the retina should be noticed, so as to compare them with the same parts when hardened.

In the posterior portion may be seen, through the transparent vitreous humor, the entrance of the optic nerve, and the blood-vessels which radiate from this point over the delicate, transparent retina, the latter spreading like a film over the whole of the interior of this half of the eye.

Remove the vitreous humor and the retina, so that the brightly colored *choroid coat* may be seen.

Within the front half of the eye will be seen a portion of the vitreous humor and the perfectly transparent lens.

The aqueous humor is best shown by taking an *entire* fresh eye, and cutting through the cornea, when this fluid will be seen to spurt out.

Both transverse and longitudinal sections should be examined. The latter are best made by placing the eye upon a piece of cork, with the cornea downwards, and then, passing the razor through the middle of the optic nerve, cut the eye in half.

When the razor is felt to be in contact with the hardened lens, cease to draw it, and finish the section by simply pressing it through, otherwise the lens will be drawn out of its place.

144. Get a sheep's kidney at the market. Study carefully its shape, size, etc., as a whole. Freeze it, if necessary, and then cut it in two lengthwise.

Notice the difference between the inner and the outer portions. This is something like the difference in texture between the pith and the hard rind of some flower-stems, such as the sunflower.

The outer part is called the *cortical* part, which means the "bark;" and the inner is called the *medullary*, or "marrow."

---

## EXPERIMENTS ON THE THROAT AND VOICE

### [CHAPTER XIV. PAGE 280.]

EXPERIMENT 145. *To show the Anatomy of the Throat.* — Study the general construction of the throat by the help of a hand-mirror. Repeat the same on the throat of some friend.

146. *To illustrate the Passage of Air through the Glottis.* — Take two strips of india rubber, and stretch them over the open end of a boy's "bean-blower," or any kind of a tube. Tie them tightly with thread, so that a chink will be left between them.

Force the air through such a tube with a bellows, or even by blowing hard, and a sound will result if the strips are not too far apart. The sound will vary in character, just as the bands are made tight or loose.

147. "A very good illustration of the action of the vocal band in the production of the voice may be given by means of a piece of bamboo or any hollow wooden tube, and a strip of rubber, about an inch or an inch and a half wide, cut from pure sheet rubber such as is used by dentists for rubber dams.

"One end of the tube is to be cut sloping in two directions, and the strip of sheet rubber is then to be wrapped round the tube, so as to leave a narrow slit terminating at the upper corners of the tube.

"By blowing into the other end of the tube the edges of the rubber bands will be set in vibration, and by touching the vibrating membrane at different points so as to check its movements it may be shown that the pitch of the note emitted depends upon the length and breadth of the vibrating portion of the vocal bands." — Dr. H. P. Bowditch.

148. *To show the Construction of the Vocal Organs.* — Get a butcher to furnish two windpipes from a sheep or a calf. They differ somewhat from the vocal organs of the human body, but they will enable us to recognize the different parts which have been described, and thus to get a good idea of the gross anatomy.

One specimen should be cut open lengthwise in the middle line in front, and the other cut in the same way from behind.

## EXPERIMENTS ON MATTERS OF EVERY-DAY HEALTH

[CHAPTER XV.   PAGE 249.]

EXPERIMENT 149. Imagine a pupil with his fingers or ears frost-bitten. Show how friction should be applied.

150. *Poultices.* — Bring a small lot of the necessary material into the schoolroom, to show exactly how a mustard paste, a flaxseed poultice, or any other simple poultice, is made. A small poultice four inches square is large enough for the experiment.

Let one or more pupils use the same material, and make a number of poultices.

151. *Fainting.* — Select several places about the schoolroom, and show exactly how a person should be placed, supposing he has fainted in a crowded room. Go through every detail of treatment in your mind.

152. *Apparent Drowning.* — Show exactly how artificial respiration is done. Let a boy lie on a settee, and illustrate the process in every detail. It would be an excellent idea for a teacher to meet his boys at their bathing-place, and show them in still more detail.

Let two boys go through the process on a playmate under the eye of the teacher, and then others might follow their example.

153. *To carry an Injured Person.* — Take some one of the small boys, and show how he could be carried home or elsewhere in case of injury.

154. *Burns.* — Have a small quantity of soda, sweet-oil, and lime-water in the schoolroom. Imagine a pupil has

burned his arm or hand.  Show exactly what is to be done, and how.

155. *Foreign Bodies.* — A thoughtful teacher can make several experiments to illustrate the text.

156. *Experiments on the Matter of treating Cuts and Bruises, and of stopping Bleeding.* — Let red-pencil marks made on the face, arms, fingers, etc., stand for cuts.  Apply suitable strips of plaster in a proper way for a great variety of imaginary cuts.  After putting on the plaster, practise on bandaging the parts with strips of cotton cloth rolled for the purpose.  Practise on using the handkerchief for a variety of bandages.

157. *To stop Bleeding from the Arteries.* — Locate the principal arteries on your own person and that of a friend. Let red-crayon marks stand for the course of the arteries. Study this part of your anatomy thoroughly.

Now with strings, cords, shoestrings, handkerchiefs, elastics, strips of clothing, practise tying them so as to press deeply and firmly in the proper place.  Let each one in the class practise at the same time on the same artery.  Criticise and improve each other's work.

158. *Poisons and their Antidotes.* — Small samples of the more common and important poisons should be shown. Note carefully the appearance of each one.  With each sample of poison, arrange its most common antidote.  Explain exactly how it should be given.

159. *Disinfectants.* — Exhibit samples of the common disinfectants, and show how they should be used.  Most school-premises furnish ample opportunities to make a series of practical experiments on ventilation and disinfection.

# REVIEW ANALYSIS

## *ALCOHOLIC DRINKS, TOBACCO, AND OTHER NARCOTICS*

---

[The figures in full-face type, in the parentheses, refer to the numbers of the section in the preceding chapters of this book.]

**1. The Bones.**
(Chap. ii. p. 13.)

Effect of Alcohol and Tobacco on the Bones **(35).**

**2. The Muscles.**
(Chap. iii. p. 42.)

1. Effect of Alcohol on the Muscles (**44**).
2. Effect of Alcohol on Strength (**45**).
3. Effect of Tobacco on the Muscles (**46**).

**3. Physical Exercise.**
(Chap. iv. p. 56.)

Effect of Alcohol and Tobacco on Physical Exercise (**58**).

**4. Food and Drink.**
(Chap. v. p. 70.)

Effect of Drinking Tea and Coffee (**72**).

**5. Origin and Nature of Fermented Drinks.**
(Chap. vi. p. 81.)

1. Change produced by Fermentation (**74**).
2. Ferments and what They do (**75**).
3. Alcohol a Poison (**76**).
4. Narcotic Poisons (**76**).
5. The Alcoholic Appetite (**77**).
6. The Evils of Cider-drinking (**78**).
7. Wine, and why it should be avoided (**79**).
8. Beer:
    Its Origin.
    Relation to Drunkenness (**80**).
9. Distilled Liquors (**81**).
10. The Cost of the Alcohol Vice. (Note, p. 89.)

**6. Digestion.**
(Chap. vii. p. 92.)

1. Effect of Alcohol on the Stomach Digestion (**99**).
2. Effect of Alcohol on the Liver (**100**).
3. Effect of Tobacco on Digestion (**101**).

329

# TEST QUESTIONS FOR REVIEW

[CHAPTER I.   PAGE I.]

1. How may the child get some idea that his body moves, and is warm?
2. What marked difference is there between all animals and all vegetables?
3. How would you compare the body to a locomotive?
4. Can you tell where the comparison fails?
5. Explain, in a general way, how we are able to cross a room, pick a flower, and do many similar acts.
6. What does the child experience when he touches a hot stove, or pricks his finger?   What good does pain do?
7. Why do we wonder at the marvellous actions of our bodies?
8. What is meant by anatomy? physiology? hygiene?   Illustrate.
9. What is an organ? a system? an apparatus?
10. What is meant by the function of an organ? Illustrate.
11. What is a vital organ?
12. What is a tissue?

13. Mention the principal tissues of the body.
14. Of what is the minute structure of the body made up?
15. Tell what you know about cells.
16. What chemical elements are found in the body?
17. What can you say about the general build of the body?
18. Mention the general topics which will be discussed in this book.

---

### THE BONY FRAMEWORK

### [CHAPTER II.   PAGE 13.]

1. What do you mean by the skeleton?   What are its uses?
2. How many bones are there in the human body?
3. Describe the composition of bone.
4. What experiments show the composition of bone?
5. Explain the general structure of bone.   Illustrate.
6. Explain and give examples of the different kinds of bones.
7. Describe the minute structure of bone.
8. What is the periosteum? and what is its use?
9. What can you say of the power of bone to resist decay?
10. Name the principal divisions of the human skeleton.
11. Into what parts is the head divided?

12. What is the cranium? Mention each bone by name.
13. Give an account of the bones of the cranium.
14. What is the face? Name each bone.
15. Describe the bones of the face.
16. How are the bones of the head joined together?
17. What are sutures? Explain fully their use.
18. What are the fontanelles?
19. What is the trunk? its two cavities?
20. What is the abdomen? and what does it contain?
21. Describe the diaphragm.
22. Of what parts does the trunk consist? Mention each bone.
23. Describe in full the spine. Its uses.
24. What are the ribs? what are their uses?
25. How are the skull and the spine joined together?
26. How are the hips made up?
27. Describe the hyoid bone.
28. Describe the upper limbs. Mention the bones of each part.
29. Describe the three bones of the upper arm.
30. Describe the two bones of the fore-arm.
31. Describe the wrist, the palm, and the fingers.
32. Describe the lower limbs. Name each bone in the three parts.
33. Describe the thigh, the lower leg, the foot.
34. What is the knee-cap? the tendon of Achilles?
35. Give some points of comparison between the hand and the foot.
36. Explain the use of bones.
37. How do bones grow? How are they repaired?

38. What is a joint? Describe in full.
39. Mention the different kinds of joints. Illustrate.
40. What are the ligaments? Describe them. What are their uses? Name some important ligaments.
41. Give some hints about the health of bones.
42. What is the effect of alcohol upon the bones?
43. How does tobacco injure the bones?
44. Give from memory the "review analysis" of the skeleton.

---

### THE MUSCLES

### [Chapter III. Page 42.]

1. How do we move? Explain in full.
2. What is muscle? Describe its structure.
3. Explain and illustrate contraction.
4. Describe the two general kinds of muscles.
5. How are muscles arranged to do their work? Illustrate.
6. What are tendons? Where are they found? What are their uses?
7. Explain how muscles act as levers of the three different classes.
8. Give examples of each class.
9. What is the number of muscles?
10. How are they adapted to different uses?
11. Mention and describe some of the important muscles of the body.
12. How does alcohol affect the control of the muscles?

13. How does alcohol affect the speech?
14. Why does alcohol sometimes give rise to the idea that it adds to or renews strength?
15. How can this idea be shown to be false?
16. Explain what is known as fatty decay.
17. What is the effect of fatty decay on the heart, liver, and other organs?
18. How do the muscles of the beer-drinker differ from those of the total abstainer?
19. How does tobacco affect the muscles? How is the effect shown?

---

### PHYSICAL EXERCISE

[CHAPTER IV. PAGE 56.]

1. What is meant by physical exercise?
2. Why do we need it?
3. What are some of the ill effects of too little exercise?
4. In a general way, describe the effect of exercise on the various organs.
5. How much exercise should we take?
6. Mention some points of contrast between health and strength.
7. What is the best time for exercise?
8. Give in detail the different kinds of exercise.
9. Show the beneficial effect of physical exercises in schools.
10. Why is a certain amount of physical training an important matter in our school system?

11. Give some points about the importance of physical development.

12. In a general way, tell what you can about the
various systems of gymnastics.

13. Give in some detail the essential features of the
Swedish system.

14. What are the important features of the German
system ?

15. What benefits may be derived from a modern
gymnasium ?

16. What system of gymnastics is needed for the
schools in this country ?

17. What features are essential for any system to be
popular and beneficial in our schools ?

.

---

## FOOD AND DRINK
### [CHAPTER V. PAGE 70.]

1. To illustrate work, waste, and repair, how may
you compare the human body to a steam-engine ?   Explain in detail.

2. How does the comparison fail in some things ?

3. Why do we need food ?

4. What are the three great classes of foods ?

5. Describe and illustrate the nitrogenous foods
(albumens).   Give examples.

6. What are the starches and sugars ?   Give examples.

7. Describe and illustrate the oils and fats.

8. What is meant by the inorganic or mineral foods ?

9. What do you mean by "appetizers"? Illustrate.
10. Describe the principal kinds of vegetable food.
11. Explain the importance of salt, and give many illustrations of its value.
12. What are some of the principal articles of animal food?
13. What is the use of mineral food?
14. What is the natural drink provided by Nature for all living things?
15. Explain how water acts to rid the body of waste and refuse matters.
16. What is essential for water to be suitable as an article of diet?
17. Mention some of the more common artificial drinks.
18. Tell what you can about tea, coffee, and cocoa.
19. What can you say of ice-water?
20. In a general way explain the harmful effects of tea and coffee.

---

## ORIGIN AND NATURE OF FERMENTED DRINKS

### [CHAPTER VI. PAGE 81.]

1. Of what are fruit juices composed?
2. How long will such juices remain healthful after they have been drawn off from the fruit and left standing in ordinary air?
3. Why does plant and animal matter decay?
4. What name is given to the process of decay that goes on in fruit juices that are pressed out and left exposed to the air?

5. What does fermentation in its widest sense include?

6. What is the law of fermentation?

7. What causes the various processes of decomposition?

8. What would be the condition of the earth if it were not for these minute living forms?

9. To what class or plant family do the disease germs belong?

10. Name other plant cells belonging to this family, and tell what they do.

11. What is the work of the moulds? To what class do they belong?

12. To what class do the ferments belong? What is their work?

13. When do the ferments enter the fruit juice?

14. Show what the ferments do in the fruit juice.

15. What is the nature of the fruit juice before fermentation? What is its nature after fermentation? Of what law is this an illustation?

16. What further changes take place in wine or cider when left standing in warm air?

17. What would become of the vinegar if left to itself?

18. What is a poison? Show why alcohol is a poison.

19. What are narcotic poisons?

20. What are the characteristics of a natural appetite?

21. What is the difference between a natural appetite and the appetite for alcohol?

22. Upon what does the character of a substance depend?

23. What is the secret of the drunkard's craving for alcohol?
24. What is the only safeguard against forming the alcoholic appetite and the only cure when it is formed?
25. Why should alcoholic liquors never be used as a flavoring for food?
26. What is cider? What poison does it contain?
27. How soon after cider comes from the press can alcohol usually be found in it?
28. Why is cider an unsafe drink?
29. What follows if a person drinks the same quantity of cider every day while it is growing hard?
30. From what is wine made?
31. Why are we in no danger of being injured by alcohol if we take our fruit juices fresh from the fruit?
32. Why are some wines called "light wines"? Why are these unsafe drinks?
33. What disproves the theory that "light wines" will prevent the use of stronger drinks?
34. From what is beer made?
35. Show how alcohol is formed in beer.
36. What effect does beer appear to have upon the moral nature?
37. Describe the beer-drinker's appearance and dangers.
38. What is the testimony of life-insurance companies in regard to beer-drinking?

39. What other drinks are sometimes made by allowing yeast to act on the sugar of a sweet liquid?
Why should no such liquors be made and used
as beverages?

40. Show how distilled liquors are obtained.

41. How is the craving for the stronger or distilled
liquors acquired?

---

### DIGESTION

[CHAPTER VII.　PAGE 92.]

1. Describe the digestive tube as a whole.

2. What is meant by digestion?　Name the various
steps.

3. What is meant by chewing?

4. Describe the structure of a tooth.　Mention the
number and situation of the teeth, giving the
names of each kind.

5. What can you tell of the salivary glands and the
saliva?　Explain what is meant by secretion
and excretion.

6. Explain in some detail how we swallow food.

7. What prevents food from getting up into the
nose, and down into the windpipe?　Explain
in full.

8. Describe the stomach.

9. What is the gastric juice?　How is it secreted?
Describe its action.

10. What are the intestines?

11. Describe the liver. The bile. The pancreatic juice. The intestinal juice. The action of each.

12. In a general way, describe the process of absorption.

13. Describe the process of absorption by means of the lacteals and the lymphatics.

14. What is the thoracic duct and its office?

15. Lymphatic glands and the work they do?

16. Explain absorption by the blood-vessels.

17. Describe the .spleen, the thymus and thyroid glands.

18. Describe the large intestines.

19. *How much* should we eat? Various circumstances affecting this question.

20. What guides have we as to *what* we should eat?

21. Describe in full *when* we should eat.

22. *How* should we eat? Explain in detail.

23. Give some practical points about the care of the bowels.

24. Why is a knowledge of proper cooking so necessary?

25. Care of the teeth. Its great importance. Give details.

26. Describe the indigestion due to alcohol.

27. What is the effect of alcohol on the stomach?

28. What is its action on the liver?

29. What is the effect of tobacco on digestion?

## THE BLOOD AND ITS CIRCULATION
### [CHAPTER VIII.   PAGE 122.]

1. Of what use is the blood ?
2. How is the blood made up ?
3. What is the total quantity in the body ?
4. How does it look ? its color ?
5. How does blood look under the microscope ?
6. Describe the red blood corpuscles.
7. Describe the white blood corpuscles.
8. What is meant by the coagulation of blood ?
9. How may you compare the flow of blood circulation to a force-pump? to the water-system of a great city ?
10. Describe the heart.
11. Explain the terms auricle, ventricle, valve.
12. What can you tell of the work done by the heart ?
13. Describe the action of the valves of the heart.
14. Name the blood-vessels connected with the heart.   Tell with which part of the heart each is connected.
15. Describe in a general way the differences between the veins and arteries.
16. Describe the aorta.   Some of its great branches
17. What are the veins ?   Describe their valves.
18. To what may the venous system be compared ?
19. What are the capillaries ? their use ?
20. Give a general description of the circulation of the blood through the body.

21. How much time does it take?
22. Describe the sounds of the heart.
23. What is meant by palpitation of the heart?
24. What is the pulse? What may it tell us?
25. What is the general effect of alcohol on the circulation?
26. How does alcohol get into the blood?
27. How may alcohol affect the corpuscles of the blood?
28. In what other way may the blood be injured by alcohol?
29. What is the effect of alcohol upon the heart?
30. What is the harm in thus increasing the frequency of the heart's beats?
31. What is the real cause of the so-called "stimulating" effect of alcohol on the heart?
32. What is the effect of tobacco on the heart?

------

### BREATHING
### [CHAPTER IX. PAGE 142.]

1. What is the object of breathing?
2. *Why* do we breathe?
3. Describe the air-passages.
4. How are they lined and protected?
5. Give a short description of the lungs.
6. What is the pleura?
7. Describe the mechanical movements of breathing.
8. *How* do we breathe?

9. Describe the changes in the air from breathing.
10. How may we compare the lungs to a market-place?
11. What is carbonic acid? where found? its poison-ous effects?
12. What impurities may exist in the air?
13. What are the chief dangers from impure air?
14. What are the practical effects of impure air?
15. Why and how should we ventilate? Give some practical plans.
16. The ventilation of schoolrooms, — why and how is it done?
17. What is the proper temperature of our living-rooms? Explain in some detail.
18. Animal heat. How is the body kept warm?
19. Explain and illustrate this slow combustion, or oxidation.
20. What is the effect of alcohol upon the lungs?
21. Explain in detail the action of alcohol upon the bodily heat.
22. Show how alcohol lessens the power to endure hardship and extremes of heat or cold. Give illustrations.

----

## HOW OUR BODIES ARE COVERED
### [CHAPTER X. PAGE 161.]

1. What is the skin? and what is its use?
2. Describe its structure in a general way.
3. Give in detail a description of the cuticle, or scarf-skin.

4. Where is the coloring-matter of the skin situated?
5. Describe the cutis, or true skin.
6. Show how the skin may absorb poison.
7. Describe the hair; its structure; muscles; number of hairs.
8. Describe the nails.
9. What care would you take of them?
10. What are the oil-glands?
11. Describe the sweat-glands.
12. What is the use of the sweat?
13. Give some reasons why we should take care of the skin.
14. Baths. Why, when, and how to take them.
15. Why do we need clothing?
16. Give some hints about the use of clothing.
17. What is the effect of alcohol upon the skin?

---

## THE NERVOUS SYSTEM

### [CHAPTER XI. PAGE 177.]

1. Show how the various parts of the body act in harmony.
2. Draw a parallel between the nervous system and the telegraph.
3. Give familiar illustrations to show the general action of the nervous system.
4. What are the two distinct systems?
5. What is nerve-tissue?

6. What are the nerves?
7. Give a short description of the brain.
8. What is the usual size of a human brain?
9. Describe fully the cerebrum, or brain proper.
10. What are the functions of the cerebrum?
11. Describe the three coverings of the brain.
12. Locate and describe the cerebellum. What is its function?
13. Describe the medulla oblongata. What are its functions?
14. What do you understand by the cranial nerves?
15. How are they arranged? What is the pneumogastric nerve?
16. Describe briefly the spinal cord. What are its functions?
17. Explain fully what you understand by reflex action.
18. Give several familiar examples to illustrate reflex action.
19. Describe the importance of reflex action.
20. What are the spinal nerves? Explain motor and sensory nerves.
21. Describe the sympathetic system of nerves.
22. What are the functions of the sympathetic nerves?
23. What tends to keep the nervous system in good order?
24. What about the abuse of the nervous system?
25. Why is sleep necessary?
26. Give some practical hints about sleep.

27. What is one of the most fruitful sources of injury to the nervous system?
28. What are narcotics?
29. Why is the person who takes alcoholic drinks once liable to take them again?
30. How does alcohol affect the substance of the nerves?
31. What causes the face to flush when a small quantity of alcohol has been taken?
32. What is the first effect of alcohol on the brain?
33. What shows that this unusual activity of the brain is an unhealthful excitement?
34. What are some of the common liquors that have this power to weaken the judgment and blunt the moral sense?
35. What part of the brain is first brought under the poisonous influence of alcohol?
36. How is this shown?
37. What part of the brain is next affected when larger quantities of alcohol are taken?
38. In what way is this shown?
39. Describe the last stage of alcohol poisoning.
40. What injurious effects follow the continual use of alcohol?
41. What terrible disease often results?
42. What is the relation of drink to insanity?
43. What are some of the other evil consequences brought upon the nerves and nervous system by the use of alcohol?
44. What evil inheritance may result from alcoholic habits?

45. What are some of the ill effects of tobacco upon the nerves?
46. How does it affect the mind?
47. Show why cigarettes are especially injurious.
48. How is the effect of tobacco on the moral nature frequently shown?
49. Why is it no act of kindness to offer a person a cigar or cigarette?
50. Opium. What is it? for what used? what about the opium habit?
51. Give some practical hints about opium.
52. What about chloral? other narcotics?
53. What new drugs are hurtful to health?

---

### THE SPECIAL SENSES

### [CHAPTER XII. PAGE 210.]

1. What is meant by the term sensation? Illustrate.
2. What is the real centre of sensation? Illustrate.
3. What are the three organisms necessary to receive sensation?
4. Give several examples to show this.
5. Name and define the five special senses.
6. What is meant by the "muscular sense"?
7. Tell how the skin acts as an organ of touch.
8. What parts of the body are most sensitive to touch?
9. Does the sense of touch become more sensitive by practice?

10. Describe the tongue as the principal organ of taste.
11. Where is the sense of smell located?
12. How are we able to detect an odor?
13. What are the uses of the sense of smell?
14. What is the sense of hearing?
15. Explain in a general way how sound travels through the air.
16. Describe the outer ear.
17. What is the middle ear? the inner ear?
18. Explain just how we are able to hear.
19. Give some practical hints about the care of the ear.
20. What is meant by the sense of sight?
21. What is the eye? where located?
22. Describe the coats of the eye.
23. Explain in some detail the inner parts of the eye.
24. Describe in full the mechanism of vision.
25. What is meant by the term "focus"? Illustrate.
26. How are images brought to a focus on the retina?
27. Show how the eye is moved by muscles.
28. What use are the eyelids? eyebrows?
29. Tell what you know of the tears, tear-glands, and ducts.
30. What is meant by color-blindness?
31. What is meant by near-sight and far-sight?
32. Give some practical hints about the care of the eyes.
33. What is the general effect of alcohol and tobacco on the special senses.
34. How do alcohol and tobacco affect the hearing?

35. What effect do they have on the eyes and eye-sight?

36. Show how cigarettes are especially hurtful to the throat and the eyes.

---

### EXCRETION
[CHAPTER XIII.   PAGE 235.]

1. Why is it necessary to get rid of the waste matters of the body?

2. How would you compare the body to an engine in this matter of getting rid of waste material?

3. Describe the principal waste matters of the body.

4. What are the three great organs of excretion?

5. How are their functions related one to the other?

6. Describe the kidneys.

7. What are their functions?

8. What is the general effect of alcohol upon the kidneys?

9. What effect does alcohol have on the tissue-waste?

10. How does alcoholic liquor change the structure of the kidneys?

11. What disease of the kidneys is apt to be induced by alcoholic indulgence?

---

### THE THROAT AND VOICE
[CHAPTER XIV.   PAGE 243.]

1. What is the throat?

2. How is it exposed to disease?

3. In what way can you get an idea of the throat?

4. What parts of the throat can you see by looking into a friend's mouth? Explain in full.
5. Describe the pharynx.
6. Give some hints about the care of the throat.
7. Describe the organ of speech.
8. What are the vocal cords? Describe them.
9. What is meant by the voice?
10. Upon what does the compass of the voice depend?
11. What regulates the quality of the voice?
12. What is the general effect of alcohol on the voice? on the vocal cords? on the delicate lining of the throat?
13. How does tobacco affect the vocal cords?
14. How is sore throat frequently caused?
15. How do cigarettes affect the throat-passages and the voice?

---

## SIMPLE MATTERS OF EVERY-DAY HEALTH

### [CHAPTER XV. PAGE 249.]

1. What is meant by "simple matters of every-day health"?
2. Mention some things of the kind likely to happen.
3. What real good may one do under such circumstances?
4. Describe in detail the process of making poultices.
5. When and how would you use them?
6. What would you do for a fainting person?

7. Convulsions, and what to do for them.
8. What is meant by asphyxia? What is the treatment?
9. Describe in detail what is to be done in apparent drowning.
10. What is sunstroke or heatstroke? What would you do?
11. How would you know that a bone was broken?
12. How would you manage with a broken bone?
13. How would you carry an injured person?
14. What is the treatment for burns and scalds?
15. What is frost-bite? What would you do for it?
16. What would you do for a foreign body in the throat? in the nose? in the ear? in the eye?
17. How would you know the difference between bleeding from an artery and bleeding from a vein?
18. On what general principles would you act to stop bleeding?
19. What would you do to stop bleeding from an artery? from a vein?
20. Can you tell the difference between blood coming from the lungs and that from the stomach? If so, how?
21. What would you do for nose-bleed? bruises? cut and torn wounds?
22. What will you say about the pernicious use of alcoholic liquors in accidents and emergencies?
23. Describe in detail how you would make a sickroom look neat and clean.

24. How would you make a sick person more comfortable ?
25. Are persons likely to swallow poisons accidentally? If so, when? Mention some poisons often accidentally swallowed.
26. What general treatment would you give for all poisons ?
27. Describe the more common acid poisons, and give their antidotes.
28. Mention the alkaline poisons. Their antidotes.
29. Mention some metallic poisons, and give antidotes.
30. Mention some narcotic poisons. Their antidotes.
31. What other things may act as poisons ?
32. How may the air be poisoned ?
33. What is meant by disinfection ?
34. In what general way would you use disinfectants?
35. Mention the three most common disinfectants.
36. Describe in detail how disinfectants should be used.

# NOTES

NOTE 1. — "Besides the danger connected with the use of
alcoholic drinks, which is common to them with other narcotic poisons, alcohol retards the growth of young cells and
prevents their proper development. Now, the bodies of all
animals are made up largely of cells, as heretofore shown,
and the cells being the living part of the animal, it is especially important that they should not be injured or badly
nourished while they are growing. So that alcohol in all its
forms is particularly injurious to young persons, as it retards
their growth and stunts both body and mind. This is the
theory of Dr. Lionel S. Beale, a celebrated microscopist and
thinker, and is quite generally accepted." — *Dr. Roger S.
Tracy.*

"On entering college (Yale), the class of '91 had a list of
thirty-eight tobacco-users, or about eighteen per cent of the
two hundred and five men. At the beginning of junior year
this percentage had been slightly increased, although eighteen of the men who had been recorded as tobacco-users
had left college for one reason or another. At the end of
senior year the record stands as follows : —

"There are seventy-seven men who never used tobacco.

"There are twenty-two men who have used it slightly at
rare intervals, of whom six have begun the practice in the
last term of senior year.

There are eighteen who have been hard smokers at different periods of the course.

" There are seventy men who have used it regularly. . . .

" In weight the non-users increased 10.4% more than the regular users, and 6.6% more than the occasional users.   In the growth of height the non-users increased twenty-four per cent more than the regular users, and fourteen per cent more than the occasional users." — *Prof. J. W. Seaver, M.D., of Yale, in the University Magazine*, June, 1891.

" Smoking prevents a healthy nutrition of the several structures of the body.   Hence comes, especially in young persons, an arrest of the growth of the body, low stature, an unhealthy supply of blood, and weak bodily powers." — *Dr. J. Copland, F.R.S., of England.*

" When the Europeans first visited New Zealand, they found the natives the most finely developed and powerful men among the islands of the Pacific.   Since the introduction of tobacco, for which these men developed a passionate liking, they have, from this cause alone, become decimated in numbers, and so reduced in stature and physical well-being as to be an altogether inferior type of men." — *New York Medical Journal.*

NOTE 2. — " Dr. Parkes made three soldiers march twenty miles a day for six days, loaded with gun, knapsack, pouch, etc.   They had the same food every day; but on two days they drank brandy and water, on two coffee, and on the other two weak beef-tea.   All three found that, though the spirit revived them for a little while, they were more tired at the end of the journey on the two days when they took brandy than on any of the other two." — *Dr. N. S. Davis.*

" There is no muscle about beer.   This is abundantly illustrated by the thousands of beer and wine drinkers, who, from twenty to twenty-five years of age, were muscular, active, capable of any reasonable endurance, with a weight of one hundred and fifty pounds, but who, after moderately retarding atomic changes, and retaining old atoms by the daily use of wine or beer, have acquired a weight of two hundred pounds or more, and have lost their muscular activity and endurance to such an extent that active exercise of twenty minutes would make them puff like a heavy horse." — *Dr. N. S. Davis.*

" It has been demonstrated on all sides, at the forge, in the workshop, in the field, on the march, in the arctic region, and in the torrid zone, in physical and in intellectual labor, that the spirit-drinker fails to cope with the temperate man." — *Dr. Willard Parker.*

" It is now well known and acknowledged by scientific men, that less muscular labor can be performed under the influence of alcohol, in whatever quantity, than without it. In the performance of great feats of strength and endurance, as in the case of Weston, the famous pedestrian, alcohol has been avoided; and in the harvest-field and the workshop, and with contestants in the ancient Roman games, the advantage has ever been with abstainers. The most conclusive tests have been in armies in severe marches, where accurate observation on a large scale has been made by intelligent medical and commanding officers. In all such tests, whether in hot or cold climates and seasons, — in Africa, India, Russia, and Canada — in our own country and everywhere, it has been shown that those soldiers who abstained from alcohol could accomplish and endure more than those who indulged in it, however moderately or freely. In emergencies, those officers who allow its use at all find that it must be given when the men have accomplished their day's work and are resting after their labor. It may then blunt the sense of fatigue and promote sleep, but, unfortunately, it lessens the power of work for the next day, and if its use becomes habitual, other mischief will be done." — *Dr. A. B. Palmer.*

" Voluntary muscular power seems to be lessened, and this is most marked when a large amount of alcohol is taken at once ; the finer combined movements are less perfectly made. Whether this is by direct action on the muscular fibres, or by the influence on the nerves, is not certain. In very large doses it paralyzes either the respiratory muscles or the nerves supplying them, and death sometimes occurs from the impairment to respiration." — *Dr. Edmund A. Parkes, F.R.S.*

NOTE 3. — " One of the first things demanded of a young man who is going into training for a boat-race is, ' Stop smoking.' And he himself, long before his body has reached its highest point of purity and development, will become conscious of the lowering and disturbing effects of smoking one inch of a mild cigar. No smoker who has ever trained severely for a race, or a game, or a fight, needs to be told that smoking reduces the tone of his system : he knows it." — *James Parton.*

" A man who, after election as a member of his college crew, should be found secretly drinking beer or smoking, would be hissed out of college." — *Dr. H. Newell Martin.*

NOTE 4. — " All that has lived must die, all that is dead must be disintegrated, dissolved, or gasified ; the elements which are the substratum of life must enter into new cycles of life. If things were otherwise the matter of organized beings would encumber the earth, and the law of the perpetuity of life would be compromised by the gradual exhaustion of its materials. One grand phenomenon presides over this vast work — the phenomenon of fermentation. . . .

" Fermentation and putrefaction only represent the first phase of the return to the atmosphere and to the soil of all that has lived. Fermentation and putrefaction give rise to substances which are still very complex, although they represent the products of decomposition of fermentable matters. When sugar ferments a large portion of it becomes gas ; but alongside of the carbonic acid gas which is formed, and which is, indeed, a partial return of the sugar to the atmosphere, new substances, such as alcohol, succinic acid, glycerine, and materials of yeast, are produced. When the flesh of animals putrefies, certain products of decomposition, also very complex, are formed, with the vapor of water and other gases of putrefaction." — " *Louis Pasteur: His Life and Labors.*"

NOTE 5. — " It has been known since very ancient times that sweet juices of fruits, more particularly of the grape, can be made to undergo certain changes, the result of which is that the juice is no longer a sweet, innocuous liquid, but possesses intoxicating properties. During the occurrence of this change the clear fluid becomes turbid, and its surface is covered by bubbles or froth. This latter phenomenon attracting special attention, the name fermentation was given to the process. We now know that the change consists in the transformation of the sweet substance sugar into other materials, of which the most abundant are alcohol, the body possessing the intoxicating properties, and carbonic acid, the evolution of which causes the frothing. The turbidity of the liquid is caused by the presence of numerous unicellular organisms, which increase rapidly by a process of budding." — *Halliburton's " Text Book of Chemical Physiology and Pathology.*"

" Thus in the destruction of that which has lived, all

reduces itself to the simultaneous action of these three great phenomena, — putrefaction, fermentation, and slow combustion. A living organism dies, animal, plant, or mineral, or the remains of the one or the other. It is exposed to the contact of the air. To the life which has quitted it, succeeds life under other forms. In the superficial part which the air can reach, the germs of the infinitely small *aérobies* (micro-organisms which cannot live without free oxygen) hatch and multiply themselves. The carbon, the hydrogen, and the nitrogen of the organic matters are transformed by the oxygen of the air, and under the influence of the life of these *aérobies*, into carbonic acid, vapor of water, and ammonia gas. As long as organic matter and air are present, these combustions will continue. While these superficial combustions are going on, fermentation and putrefaction are doing their work in the interior of the mass by the developed germs of the *anaérobies*, which not only do not require oxygen for their life, but which oxygen actually kills. Little by little, at length, by this work of fermentation and slow combustion, the phenomenon is accomplished. Whether in the free atmosphere or under the earth, which is always more or less impregnated with air, all animal and vegetable matters end by disappearing. To arrest these phenomena, an extremely low temperature is required. It is thus that in the ice of the Polar region antediluvian elephants have been found perfectly intact. The microscopic organisms could not live in so cold a temperature. These facts still further strengthen all the new ideas as to the important part performed by these infinitely small organisms, which are, in fact, the masters of the world. If we could suppress their work, which is always going on, the surface of the globe, encumbered with organic matter, would soon become uninhabitable." — "*Louis Pasteur: His Life and Labors.*"

NOTE 6. — " It is entirely safe to say that there is not an occupation or condition in human society in which those who use any variety of fermented or distilled liquors, even in the most 'decent and orderly manner,' do not furnish more cases of sickness and more deaths annually than are furnished by an equal number occupying the same conditions but totally abstaining from all drinks." — *Dr. N. S. Davis.*

"Alcohol taken into the body paralyzes and weakens the nerves, hardens and contracts the animal fibre; the capillaries, arteries, veins, lymphatics, or other canals and ducts for conveying fluids, are lessened in their diameter, and are ultimately obstructed, so that the foundation is laid for many diseases." — *Dr. John Higginbottom, F.R.S.*

Dr. Henry Thompson, an English physician of large practice, declares that he has " no hesitation in attributing a very large proportion of the most painful and dangerous maladies to the ordinary and daily use of fermented drinks in quantities conventionally deemed moderate."

" I state unhesitatingly that from experience and observation, the use of light wines and beer is positively injurious to many persons; and I am forced to believe that even if in others no pernicious effects be observed immediately, such effects will become apparent sooner or later."— *Dr. T. F. Allen of New York City.*

" I would say that both theoretically and practically, I do not believe in even a moderate use of light wines and beer as a beverage. The harm they do will not be manifested usually during perfect health; but in times of severe sickness I believe a person's power to resist disease is in proportion to his abstinence from all forms of liquor, even in the light forms." — *Dr. B. N. Bridgeman.*

" In addition to the injurious effects morally, which the moderate use of alcohol produces, and all drunkenness is

born of it, it is injurious physically also I believe. and chiefly
so in the following ways : (*a*) alcohol **acts detrimentally upon
the blood corpuscles** and fibrinous elements of **the blood**, producing a **ragged or cog-wheel** margin of **the former, and**
ready coagulation **of the latter**, with hardening of the coats
and consequent loss of elasticity of the blood-vessels, thus
producing the most favorable condition for plugging of the
capillaries. (*b*) By its action upon the mucous membrane **of**
the alimentary **canal and** other **portions** of the body, the
system is **less** amenable **to remedies and constant repair.**
(*c*) It increases the tendency to congestion and inflammation
generally, and in many cases produces **a fatal** result in **a not**
necessarily fatal **disease.** (*d*) What is **true** in health is even
**more true** in disease, **and** I believe that even **more cases of**
pneumonia succumb **through its use that** might otherwise
recover." — ***Dr. D. G. Dowkoutt.***

NOTE 7. — " I **can see** nothing in the action of alcohol **in**
the human body, **in any case or** at any time, but that **of a**
paralyzer ; **and I see in** that view the **key** by which we can
explain **all** the contradictory phenomena and all the contradictory benefits which have been ascribed to the influence **of**
alcohol." — *Dr. James Edmunds.*

" Alcohol, the rectified product of the vinous fermentation
(i. e., decomposition) of various saccharine fluids, and included by chemists among the narcotic poisons, exercises a
metamorphosic effect on every organ of the human body." —
*Dr. Felix Oswald.*

" Alcohol, for all intents and purposes, may be regarded
**as a sedative or narcotic** rather than a stimulant." — *Dr.
Samuel Wilks.*

" The whole class of alcoholic liquors may be considered
**as** narcotics, producing very little difference in their ultimate
effects on the system." — *Dr. William Beaumont.*

" The flushing of the face (produced by alcohol) is caused by the paralysis of the cervical branch of the sympathetic nerve. This symptom usually occurs some time before the conspicuous manifestation of the ordinary signs of intoxication which result in paralysis of the cerebrum ; we may search in vain among the phenomena of intoxication for any genuine evidence of that heightened mental activity which is said to be followed by depressive recoil. There is no recoil, there is no stimulation. There is nothing but paralytic disorder from the moment narcosis begins." — *Prof. John Fiske, of Cambridge, Mass.*

" It is the diminution of nerve sensibility that renders the individual at first light, airy, and hilarious, giving the popular idea of excitement or stimulation ; second, dull, hesitating, or incoherent in thought or speech, and unsteady or swaggering in gait, a stage popularly recognized as incipient intoxication ; and third, brings on entire unconsciousness and muscular paralysis, constituting dead drunkenness or complete anæsthesia. These successive stages are developed in direct ratio to the quantity taken." — *Dr. N. S. Davis.*

NOTE 8. — " The habitual use of alcoholic liquids, by the anæsthetic effect of alcohol on the human system, tends constantly to create an appetite for more, and consequently moderation from the beginning very generally leads directly to excess sooner or later." — *Dr. N. S. Davis.*

" I observed as a physiological or, perhaps, psychological fact, that the attraction of alcohol for itself is cumulative : that so long as it is present in the human body, even in small quantities, the longing for it, the sense of requirement for it, is present, and that as the amount of it insidiously increases, so does the desire." — *Dr. B. W. Richardson.*

" Alcohol creates an appetite for itself that soon becomes irresistible, and every drunkard will admit that when he began to drink he had no intention of becoming a drunkard." — *Dr. R. Gowans.*

" The physician who prescribes narcotics is dogged by the ever-present danger that his patients may acquire a craving for the drugs thus ordered. . . . It is a common experience that a morphia or a chloral habit owes its origin to the orders of a medical attendant, who has therefore incurred a most serious responsibility. Alcohol, as the temperance reformers have so long insisted, must be included in the list." — *The Birmingham Medical Review.*

" In the case of moderate drinking, either of malt or spirituous liquors, there is small hope that the habitual drinker will remain a moderate one." — *A. G. Bullock, President Mutual Life Insurance Company.*

" The power of moral resistance is weakened with every repetition of the poison dose." — *Dr. Felix Oswald.*

" It is the temperate use of alcohol that creates the appetite of the inebriate." — *Dr. William Hargreaves.*

" It will be admitted, I presume, by all who hear me that if there were no temperate drinking there would be none that is intemperate. Men do not begin by what is usually called immoderate indulgence, but by that which they regard as moderate. Gradually and insensibly their draughts are increased until the functions of life are permanently disturbed, the system becomes inflamed, and there is that morbid appetite which will hardly brook restraint, and the indulgence of which is sottish intemperance. Let it be remembered, then, that what is usually styled temperate drinking stands as the condition precedent of that which is intemperate. Discontinue one and the other becomes impossible." — *From an address by A. Potter, D.D., late Bishop of the Diocese of Pennsylvania.*

NOTE 9. — "Of 623 moderate and immoderate drinkers with whom I have conversed, 337 tell me that they acquired the desire for wine and other alcoholic poisons by their use in articles of diet, and in the family and social circles, dealt out to them by their wives, and sisters, and female friends. Of this number, 161 cases (more than twenty-five per cent) were from the use of liquors in articles of diet. Of the whole number referred to, 328 fill a drunkard's grave." — *A physician prominently connected with a leading life-insurance company.*

NOTE 10. — "We have been told in America, and I fully believed it, that if a people could be supplied with a cheap wine, they would not get drunk — that the natural desire for some sort of stimulant would be gratified in a way that would be not only harmless to morals, but conducive to health. I am thoroughly undeceived. The people drink their white wine here to drunkenness. . . . So this question is settled in my mind. Cheap wine is not the cure of intemperance. The people here are just as intemperate as they are in America ; and, what is more, there is no public sentiment that checks intemperance in the least. The wine is fed freely to children, and by all classes is regarded as a perfectly legitimate drink. . . . I firmly believe that the wines of Switzerland are of no use except to keep out whiskey, and the advantages of the wine over the whiskey are not very obvious. It is the testimony of the best men in Switzerland — those who have the highest good of the people at heart — that the increased growth of the grape has been steadily and correspondingly attended by the increase of drunkenness. They lament the planting of a new vineyard as we at home lament the opening of a new grog-shop. They expect no good of it to anybody. They know and deeply feel that the whole wine-producing enterprise is charged with degradation for their country." — *Dr. J. G. Holland, in the Springfield Republican.*

" That wine will intoxicate, does intoxicate, that there are confirmed drunkards in Paris and throughout France, is notorious and undeniable. You can hardly open a French newspaper that does not contain some account of a robbery perpetrated upon some person stupefied by over-drinking, — a police case growing out of a quarrel over the wine-cup, or a culprit, when asked why the sentence of law should not be pronounced on him, replying, 'I was drunk when this happened, and know nothing of the matter.' That journeymen are commonly less fitted for and less inclined to work on Monday than on other days of the week is as notorious here as it ever was in any other rum-drinking city." — *Horace Greeley, in a letter from Paris.*

" In September, 1887, the French Minister of Finance organized an alcoholic commission of which M. Leon Say was made president, to investigate the subject of alcoholism. In his report, which has since been published, he deals with the hygienic aspects, and represents that alcoholism is to-day in France one of the most serious dangers of the times. Not only men, but women and children, are affected. Mental diseases hitherto unknown have become common. A previous commission, appointed by the French Senate in 1886, reported that alcoholism threatened the people of France with rapid degeneration." — *Dr. K. Mitchell.*

" I had entertained a hope that the manufacture of pure wines and their introduction into general use would crowd out the gross strong liquors and diminish intemperance. I am now fully convinced that this hope was groundless and delusive. In wine-growing districts intemperance is on the increase, extending to the youth of both sexes." — *Rev. Dr. Stone of California, formerly of Park-street Church, Boston.*

The editor of the *Pacific* says, "Wine has become as cheap as milk, and is as freely drunk, till many once sober men are growing habitually intoxicated. One leading man enumerated to us five of his acquaintances who, once noble men, are now drunkards through wine."

"A few weeks ago there arrived at my house a niece of mine from the missionaries in Persia, born in that vine-growing land and familiar with the habits of the people. She repeats the testimony given by her father in 1867 at our State House, before the License Committee. This young lady speaks of the lamentable ravages of intemperance in that land at the present time, caused by using the wine made from the pure juice of the grape. They have large earthen jars, one-third sunk in the ground, and so high that a man must stand on a stool to reach the top. These are filled with grape-juice; sheepskin is stretched over the top and plastered with clay. After some two months it is fermented; but as it will turn to vinegar in a few weeks if opened, a man invites his friends, and for a week or ten days, till his jar is empty, they continue in a state of beastly intoxication. After a time a neighbor opens one of his jars, and a similar scene is enacted." — *Rev. A. H. Plumb, D.D., in an address given in Tremont Temple*, March, 1881.

NOTE 11. — "A copious beer-drinker is all one vital part. He wears his heart on his sleeve, bare to a death-wound even from a rusty nail or the claw of a cat." — *Dr. Grinrod, a prominent London physician.*

John S. Ford, writing from Milwaukee, Wis., to *The Boston Traveller*, says, "Physicians of this city who have had wide experience are of the opinion that the person who uses beer habitually is more liable to contract disease and less liable to throw it off than one who abstains from its use. A case recently occurred of a German brewer, apparently healthy and robust, who accidentally stuck a small sliver in his hand. Soon his arm began to swell and be painful, and the pain and swelling extended to the entire body, and resulted in death. The symptoms were clearly those of a bad condition of the blood, and no other explanation could be given than poisoning by the use of lager beer."

NOTE 12. — Prof. H. A. Hare, of the University of Pennsylvania, made seventeen experiments, and every one of them very clearly showed that the beer retarded both the salivary and gastric digestion.

Prof. Duggan, of Johns Hopkins University, found that alcohol in any form retarded the digestion of starch in a marked degree.

Prof. Paul Bert ascertained that even small doses of spirits delayed digestion for the first two hours till absorption of the liquor had been mostly effected.

When Dr. Beaumont was experimenting on Alexis St. Martin he gave him a glass of gin, when the digestive process was at once arrested, and did not begin again until after the absorption of the spirit, after which it was slowly renewed and tardily finished.

"It is an entirely false idea that alcohol acts as an aid to digestion." — *Dr. B. W. Richardson.*

" It is commonly thought that alcoholic drinks act as aids to digestion. In reality, it would appear that the contrary is the case. Any one may make the observation on himself that a meal without alcohol is more quickly followed by hunger than when alcohol is taken. The inhibitory influence of alcohol on digestion has been observed on a patient with a gastric fistula, on several other persons by the aid of the stomach pump, and by means of numerous other experiments." — *Prof. G. Bunge, of the University of Basle, Switzerland.*

"The moderate use of strong drinks is always unhealthy, even when the body is in a healthy condition. It does not do any good to the digestion, but even interferes with that process; for strong drink can only temporarily increase the feeling of hunger, but not in favor of digestion; after which strong reaction must follow, and evils which are usually attributed to other causes often result from their habitual use with moderate drinkers." — *Six hundred physicians of the Netherlands.*

NOTE 13. — "The organ of the body which most frequently undergoes structural change from alcohol is the liver. The capacity of this organ for holding active substances in its cellular parts is one of its marked physiological distinctions. In instances of poisoning by arsenic, antimony, strychnine, and other poisonous compounds, we find, in conducting our analyses, the liver to be, as it were, the central depot of the foreign matter. It is practically the same in poisoning with alcohol. The liver of the confirmed alcoholic is probably never free from the influence of the poison; it is too often saturated with it.

The effect of the alcohol upon the liver is through the minute membranous or capsular structure of the organ upon which it acts to prevent the proper dialysis and free secretion. The organ at first becomes large from the distention of its vessels, the surcharge of fluid matter, and the thickening of tissue. After a time there follow contraction of membrane and slow shrinking of the whole mass of the organ in its cellular parts, owing to the obstruction offered to the returning blood by the veins, and death is certain." — *Dr. B. W. Richardson.*

" The first effect of alcohol on the liver is to irritate it, just as it irritates the mouth and stomach, or, when applied strong enough, the skin. It causes distention of the blood-vessels and an accumulation of a larger amount of blood in them than there should be. This results in a swelling of the organ, partly from the larger quantity of blood in the vessels, and partly from effusion into it and an increase of tissue. This change in the liver causes a change in its action ; and its important work of preparing the food carried to it and making it ready for the uses of the body, its office of making blood, of changing waste matter so that it can be carried out of the system by other organs, and its work of secreting bile, are all imperfectly done.

" This defective work leads to general derangement of the whole system. There is what is called biliousness — disturbance of the stomach, a coated tongue, foul breath, deranged bowels, headache, dizziness, dimness of sight, distressing dreams, a feeling in the side and stomach, and general uncomfortable sensations. Notwithstanding that these unpleasant effects are so frequently produced by what are regarded as moderate quantities of wine, beer, or spirits, yet each drink, by its narcotic or soothing effect upon the brain and nerves, may make the person feel better for the time, just as the distress produced by opium-eating is temporarily relieved by repeating the dose.

" But much more serious effects are sometimes produced by alcoholics ; and beer is more apt to act in the way about to be mentioned than whiskey. An accumulation of fat is often produced in the liver, causing its greater and more permanent enlargement, and impairing more permanently its action. When this is the case, stopping the drink does not produce the same rapid improvement as in the cases before mentioned. But where the fat is deposited between the proper liver cells,

or structures, without taking the place of them, abstaining
from drink may in time be followed by much improvement.
There is another fatty change much worse than this where
particles of fat take the place of the structure. This is
called fatty degeneration, and where it occurs other organs
are likely to be affected in a similar way; and this disease
before a great while ends in death. When any portion of the
liver tissue is changed into fat, that part cannot do its work,
and as that change goes on action will cease and death will
follow." — *Dr. A. B. Palmer, late Dean of the Medical Department of the Michigan University.*

NOTE 14. — " It (tobacco) lessens the natural appetite,
more or less impairs digestion, and induces constipation,
while it irritates the mouth and throat, rendering it habitually
congested, and destroying the purity of the voice. It induces an habitual sense of uneasiness and nervousness, with
epigastric sinking, or tension, palpitation, hypochondriasis,
neuralgia." — *Prof. Alfred Stillé, Professor of Theory and
Practice of Medicine in the University of Pennsylvania.*

" Tobacco gives rise to debility of the stomach and nausea." — *Dr. B. W. Richardson.*

NOTE 15. — " Alcohol even in small quantities acts on the
nerve pabulum in the blood, preventing its taking up oxygen
and exhaling carbonic acid." — *Dr. George Harley, F.R.S.*

" Alcohol stimulates the blood discs to an increased and
unnatural contraction, which hurries them to the last stage of
development — that is, induces their premature decay and
death. The coloring matter is dissolved out of them and the
pale discs lose their vitality, whence less oxygen can be
absorbed." — *Prof. Shultz.*

" In the ordinary use of alcoholic drinks enough alcohol is

not taken to produce death by coagulation of the albumen of the blood, but this affords no warrant for assuming that the lesser quantity is neutral or inactive. . . .   Just to the extent in which it is present it must exert an unhealthy, abnormal influence upon the albumen." — *Prof. E. L. Youmans.*

"With all parts of the blood, with the water, fibrine, albumen, salts, fatty matter, and corpuscles, the alcohol comes in contact when it enters the blood, and, if it be in sufficient quantity, it produces disturbing action.   I have watched this disturbance very carefully on the blood corpuscles; for in some animals we can see these floating along during life, and we can also observe them from men who are under the influence of alcohol by removing a speck of blood and examining it with the microscope.   The action of alcohol, when it is observable, is varied.   It may cause the corpuscles to run too closely together and to adhere in rolls; it may modify the outline, making the clear-defined, smooth, outer edge irregular or crenate, or even star-like; it may change the round corpuscles into the oval form, or, in very extreme cases, it may produce what I may call a truncated form of corpuscles, in which the change is so great that if we did not trace it through all its stages, we should be puzzled to know whether the object looked at were indeed a blood-cell.   All these changes are due to the action of the spirit on the water contained in the corpuscles, — upon the capacity of the spirit to extract water from them.   During every stage of modification of corpuscle thus described, their function to absorb and fix gases is impaired, and when the aggregation of cells in masses is great, other difficulties arise, for the cells united together pass less easily than they should through the minute vessels of the lungs and of the general circulation and impede the current, by which local injury is produced."
— *Dr. B. W. Richardson.*

Note 16. — " Within the last few years experiments of the most exact and conclusive character have been made by skilled investigators, to determine the action of alcohol on the hearts of animals. . . . Among the most careful and skilful experiments on this subject are those of Drs. Sidney Ringer and Harrington Gainsbury of London. . . . These experiments made upon the hearts of frogs were instituted for the purpose of determining the comparative effects of the different alcohols in their direct action upon that organ. It was found that all the alcohols (including common alcohol, the active principle in all our liquors) diminished the force of the heart's action, and arrested it in a longer or shorter time, in exact proportion to the strength of the respective articles and the quantity applied. A long series of experiments furnished the same results and demonstrated their correctness. Common alcohol is weaker and lighter than some of the other rarer alcohols, but heavier and stronger than others ; but the effect in character was the same in all, differing only in degree. These eminent experimenters, in closing the report on these articles, declared that, ' by their direct action upon the cardiac tissue, these drugs are clearly paralyzants ' (and not stimulating), and that this appears to be the case from the outset, no stage of increased force or contraction preceding.

Prof. Martin, of Johns Hopkins University, who has written an excellent work on physiology, and who stands among the very highest in this country as an experimental physiologist, has made experiments with the view of determining the precise effect of common alcohol, when in the blood in certain proportions, upon that organ. The results of these experiments have not been contradicted by any other experiments of a similar character, and they conclusively prove that the direct action of alcohol on the heart is paralyzing and not stimulating.

"It is true that alcohol often, indeed generally, increases the frequency of the heart's action, but not its force, when in a previously healthy state ; except perhaps in cases where it excites feverishness, which is a diseased condition. Great frequency of the pulse is an evidence of weakness rather than of strength." — *Dr. A. B. Palmer.*

NOTE 17. — "Tobacco-smoking often causes an intermittent pulse. Out of eighty-one smokers examined, twenty-three presented an intermittent pulse, independent of any cardiac lesion. This intermittency disappeared when the habit of smoking was abandoned. Among children from nine to fifteen years of age, smoking undoubtedly caused palpitation, intermittent pulse, and chloro-anæma.

The irregularity of the heart's action, which tobacco causes, is one of its most conspicuous effects. Candidates are annually rejected for cardiac disturbances, who have subsequently admitted the use of tobacco, and the annual physical examination of cadets reveals a large number of irritable hearts (tobacco hearts) among boys, who had no such trouble when they entered school. Among the applicants for enlistment as apprentices in the navy during the year 1879, ten in a thousand were rejected for functional lesions of the heart, indicating tobacco poisoning."

Dr. Frangel, of Berlin, says that in his country, "while even immoderate smoking may appear to agree with persons for many years, suddenly, and without any other assignable cause, trouble with the heart begins, and a physician is consulted. These troubles seldom begin until the smoker has passed his thirteenth year, and usually appear at an age between fifty and sixty."

# APPENDIX

**Value of Physical Exercises.** — In preceding chapters [Chapters III. and IV.] of this book we have
discussed the importance of physical exercises, and
have alluded to the value of light
gymnastics as cheap and convenient means of giving muscular
strength and vigor to all parts of
the body.

It was claimed that children
should be trained every day at
home or at school in the use of
free gymnastics, light wooden
dumb-bells, rings, and so on. We
learned that a daily exercise of this
kind would do much to develop
feeble and narrow chests, and to check the tendency
to curvature of the spine and round shoulders so common with school children.

The study of physiology should do much to arouse
the attention of teachers and pupils to the fact that
physical culture is important and useful. The preceding sections of this book, therefore, can be utilized to
impress upon the minds of young people the value of
some kind of gymnastic training.

For the benefit of those teachers who may wish to encourage their pupils to take an interest in the matter of physical culture, the following simple gymnastic exercises have been arranged.    They are merely introductory to more extended exercises.    They may be easily abridged, extended, or modified as may be thought best.    These exercises are not new or novel. They have been carefully selected and thoroughly tested in the schoolroom, and can be practised by any pupil in ordinary health without any fear of injury.

**Hints for Practice.** — Gymnastic exercises should be practised with regularity and moderation.    An exercise of fifteen or twenty minutes, once or twice a day, is enough.    The various movements should be executed steadily and gradually and never in a spasmodic sort of way.    They should be performed with attention, force of will, and energy.    Slowly at first, after a time, practice will give vigor and precision.

Avoid all exposure and draughts between and after the exercises.    A wholesome fatigue may be felt, but not weariness or exhaustion.    Let the practice be discontinued for a time if it results in dizziness, nausea, and pains in the back or side.

Do not practise too soon after eating a full meal, but rather one hour or more before a meal.    Do not have the clothing too tight about the neck, chest, or abdomen during the exercise.

Note. — These exercises have been compiled and rearranged from Lucy B. Hunt's " Handbook of Light Gymnastics " [price 50 cents], a most excellent and practical maunal for teachers.    With the aid of this excellent little book, the teacher can easily arrange more extended and complicated exercises, suited to the special need of his own school or class.

## EXERCISES

### I. FREE GYMNASTICS.

**Position.** — Stand with heels together, hips and shoulders back, hands firmly closed and well back upon the chest.

**Directions.** — Each number fills a strain of music except when otherwise specified.

Keep the heels together and hips back, unless the exercise otherwise directs. The arms overhead should always be with elbows unbent.

These exercises should be taken slowly and with caution at first. As the strength increases, greater rapidity and force should be employed.

Music for the free gymnastics should be either in galop or polka time.

### EXERCISE I.

1. Thrust right hand down twice, left twice, alternately twice, together twice.

2. Repeat No. 1, only thrust hands out at sides instead of down.

3. Repeat No. 1, thrusting hands directly up.

4. Repeat No. 1, thrusting hands from shoulders directly forward.

### EXERCISE 2.

5. Right hand down once, left once, then clap hands through rest of the strain.

6. Same exercise, out at sides.

7. Same exercise, directly up.

8. Same exercise, out in front.

## EXERCISE 3.

**9.** Hands on the hips, step with right foot forward, then diagonally forward, directly at side, diagonally

FIG. 98.          FIG. 99.

back, directly back, cross back of left, cross again still farther back ; lastly cross in front of left foot, return· ing to position after each step.

**10.** Repeat No. 9, with left foot.

### EXERCISE 4.

**11.** Stamp with right foot forward three times, advancing each time, then left three times. Stamp three times back with right foot, same with left.

**12.** Repeat No. 11.

## EXERCISE 5.

**13.** Hands still on hips twist body alternately to right and left, twice each ; four beats of music.

**14.** Bend body alternately to right and left, four beats of music finishing the strain.

### EXERCISE 6.

**15.** Bend body alternately forward and back, twice each.

**16.** Bend body first right, then back, left, front ; reverse, left, back, right, front, finishing the strain.

Fig. 100.

**17.** Same as No. 13, only twist the head.

**18.** Same as No. 14, only bend the head instead of the body.

## EXERCISE 7.

**19.** Same as No. 15, with head only.

**20.** Like No. 16, bend head instead of body, right back, left, front, then reverse.

## EXERCISE 8.

**21.** Arms extended in front, bring them forcibly back to chest eight times.

**22.** Arms again extended, raise right hand twice without bending the elbow, then left twice, alternately twice, together twice.

### EXERCISE 9.

**23.** Hands closed on chest, thrust down, out, up, and in front, twisting the arms each thrust ; repeat.

**24.** Thrust hands from chest toward floor without bending the knees, stop on chest, then over head, rising on toes, and opening hands at each thrust, continue in half time through strain.

**25.** Cross left foot over right, at same time touching fingers over head ; then right foot over left, alternately in half time through the strain.

FIG. 101.

### EXERCISE 10.

**26.** Stamp left foot, then right, charge diagonally forward with right foot, bend and straighten right knee, at the same time carrying arms back from horizontal in front. When the arms are extended in front, the hands should be the width of the shoulders apart.

**27.** Repeat this exercise on the left side.

---

#### II. EXERCISES WITH DUMB-BELLS.

**Position.** — Heels together, hips and shoulders back, bells down at sides. One-half of each strain of music is given to the exercise, the other half to what is called "the attitude." In taking these attitudes the

bells are brought first to the chest ; then, unless otherwise specified, placed upon the hips.

**Directions.** — Step carefully but quickly to all the
attitudes.

Rest oftener than in the other exercises.

Use too light rather than too heavy dumb-bells.
Old-fashioned waltzes, like the " Boston Dip," are best
for these exercises. Scotch airs,
and airs from popular operas in
this time, can easily be adapted
by a skilful musician.

### EXERCISE II.

**28.** Hands down at sides, palms
in front, turn bells four times,
bringing them to chest on fourth
accented beat.

*Attitude :* Step diagonally for-
ward with right foot, carrying hands to hips, looking
over right shoulder.

**29.** Elbows at sides, turn bells just half-way round
four times.

*Attitude :* Step diagonally forward with left foot,
looking over left shoulder.

**30.** Arms extended at sides, turn bells four times.

*Attitude :* Step diagonally back with right foot, looking over right shoulder.

**31.** Arms extended over head, palms in front, turn
bells four times.

*Attitude :* Step diagonally back with left foot, looking
over left shoulder.

EXERCISE 12.

**32.** Bells far back on chest, thrust both down, out at sides, up, and out in front.

*Attitude:* Turn to the right, throw arms up at side without bending the knees. The bells in this attitude should be exactly horizontal and parallel.

**33.** Repeat No. 32, turning to the left and throwing the arms up on left side.

*Attitude:* Repeat attitude No. 32.

FIG. 102.

EXERCISE 13.

**34.** Drop bells at sides, right hand up to armpit once, left once, together twice.

*Attitude:* Drop to sitting position, bells touching the floor, rest through the remainder of the strain.

EXERCISE 14.

**35.** Bells on shoulders, thrust each up once, both together twice.

*Attitude:* Rise on toes, palms forward, bells parallel.

**36.** Arms extended in front, turn four times.

*Attitude:* Step diagonally forward with right foot, right hand on hip, looking back at left bell, which is extended in left hand.

### EXERCISE 15.

37. Arms extended sideways at an angle of forty-five degrees, turn bells four times.

*Attitude:* Step forward with left foot, left hand on hip, looking back at right bell, which is extended in right hand.

### EXERCISE 16.

38. Bells on chest, right hand down, then up, left hand the same.

*Attitude:* Turn body to right, thrust right hand obliquely up, palm up; left hand obliquely down, palm down.

### EXERCISE 17.

39. Bells on chest, right hand up, left down; reverse, then both down, both up.

*Attitude:* Turn to left, thrust hands up and down as in No. 38.

### EXERCISE 18.

40. Arms extended in front, palms opposite, right hand up once, left the same, both together up twice.

This should he done without bending the elbows.

*Attitude:* Step diagonally forward with right foot, the body and head thrown forward, and arms thrown wide apart.

41. Repeat No. 40.

*Attitude:* Repeat attitude No. 40, on the left side.

### EXERCISE 19.

42. Arms extended at sides, right arm up once, left once, both twice, without bending the knees.

*Attitude :* Step diagonally back with right foot, right hand up, with bell perpendicular, left hand on hip.

**43.** Repeat No. 42.

*Attitude :* Repeat attitude on left side.

### EXERCISE 20.

**44.** Arms extended, with bells parallel in front, bring the bells back forcibly upon the chest four times.

*Attitude :* Fold the arms with bells closely pressed against the chest, and bend back slowly from the waist.

---

### III. EXERCISE WITH WANDS.

**Directions.** — These exercises are performed in couples, partners facing each other about three feet apart ; the one standing on right of teacher on platform holding both rings.

Schottische time is the best, but slow marches and quicksteps can be used.

Always select a wand just long enough to reach the armpit when placed on the floor at one's side. All exercises from behind the head or back should be taken with caution, and avoided altogether by those with weak backs.

**Note.** — For several of the illustrations in this chapter, the author is indebted to the publishers (Educational Publishing Company, Boston), for kind permission to take them from their excellent book called " Ladies' Home Calisthenics."

**Position.** — Heels together, hips and shoulders well back. The wand is held in front of the right shoulder, till first signal from the piano, which consists of three chords struck with both hands, the first being the length of the other two ; then drop it horizontally in front of the body. At second signal raise the wand till the arms are extended in horizontal position in front of body, place the hands so as to divide the wand into three equal parts. At third signal, carry the wand back to second position down in front.

The simplest of Strauss's waltzes must be used, or those of other composers similar in style.

**Directions.** — In all exercises, turning back to back, be careful and not pull suddenly, and never let go the ring before the word is given.

FIG. 103.

Always stand at such a distance from next couple that there can be no hitting of rings.

The rings should always be strongly made, and about six inches in diameter.

**45.** Raise the wand to chin four times, keeping elbows high, last time carry it above the head, then bring down under chin four times.

**46.** Carry wand from above the head nearly to floor, four times, without bending knees or elbows, then down back of the neck four times.

**47.** Carry wand from above the head to chin, and then back of neck, alternately four times each.

**48.** Wand over head. On first beat, carry right hand to right end of wand, on second beat, left hand to left end, then carry hand back of head to hips, six times, keeping elbows stiff.

**49.** Carry wand back from above head down nearly to floor; and then back to hips, four times, alternately four times each.

**50.** Carry wand from above the head to right and left sides alternately eight times, keeping elbows stiff, and stopping exactly over head each time.

**51.** On first beat, let go wand with left hand, place end of wand on floor between feet. On second beat place wand on floor at arm's length, diagonally forward on right side. Step with right foot to wand through rest of strain, keeping right arm, left knee, and wand perfectly straight.

**52.** Repeat No. 51 on left side.

53. Repeat No. 51, keeping the foot stationary, the knee bending with each accented beat.

54. Repeat No. 53 on left side.

## EXERCISE 24.

55. Arms horizontal in front, wand held perpendicularly, bring wand back to chest eight times, keeping elbows high.

56. Wand and arms in same position, bring wand to right and left shoulders alternately four times each. In passing the wand from one side to the other, raise the arms straight to a horizontal position in front.

## EXERCISE 25.

57. Hands in front of chest, point wand diagonally forward at an angle of forty-five degrees, first to the right, then to the left alternately through strain, making the change of hands just in front of chin.

58. With wand pointing in the same direction as in last exercise, step diagonally forward with right and left foot alternately through strain.

FIG. 104.

**59.** Repeat No. 58, only step back **instead of forward,** leading **with** left foot instead **of** right, keeping **wand** pointing forward.

## EXERCISE 26.

**60.** Wand horizontal **over head,** right hand in front, reverse position, bringing left hand in front, on half time through strain.

61. Same position, right face, bend forward bringing wand to perpendicular on right side, four times.

**62.** Repeat No. 61 **on** left side.

## EXERCISE 27.

**63.** On first beat, put left end of wand on floor **in front** of feet; on second **beat,** carry wand at arm's length in front, charge **right foot to** wand **twice, left** four times, changing hands and feet at same time.

64. Right foot back four **times, right** hand on **wand,** same with left hand and **foot.**

**65.** Right foot forward and back four times, left the same, holding wand in same position as last exercise.

**66.** Both **hands** on wand in front, right foot forward, **left back** at the same time, reverse and repeat.

----

## IV. EXERCISES WITH RINGS.

## EXERCISE 28.

67. On **first beat** of music, the ring in right **hand is** extended, and grasped by partner's right hand. Second **beat,** right feet together, toes touching; on **third** beat left feet back at right angles with right feet, with left

hands upon hips. Turn the ring over half-way and then back to place through rest of strain, keeping perfect time.

68. Repeat No. 67, only use left hand and left foot instead of right.

69. Repeat No 67, only first join both hands, on second beat right feet together, third beat step back as before, turn rings through strain.

70. Repeat No 69, with both hands joined and left feet touching, right feet back, turn rings through strain.

### EXERCISE 29.

71. On first beat, turn back to back, on second beat left feet together, charge directly forward with right feet ; head and shoulders well thrown back, pull evenly with partner, and turn the rings through strain.

72. Repeat No. 71, with right feet together, left out in front, turn rings through strain.

### EXERCISE 30.

73. On first beat, turn face to face, on second beat raise arms above head, then lower rings without bending knees, looking alternately to right and left of partner through strain.

74. First beat, lift arms towards platform, high up at side, the others low down at the opposite side, carry them alternately up and down through half the strain, then both together, half a strain.

## EXERCISE 31.

**75.** First beat, turn back to back, charge diagonally forward with right and left feet alternately through strain.

**76.** First beat, turn face to face, place left foot inside partner's left, short step back with right foot at right angles with the left. Rings over head held firmly, arms perfectly straight, sway alternately through the strain.

**77.** Repeat No 76, with right feet together instead of left.

## EXERCISE 32.

**78.** First beat, turn back to back, charge up and down the hall alternately twice each ; charge with right feet at same time, then left feet at same time, alternately through rest of strain.

**79.** First beat, turn face to face, repeat No. 78.

BOOKS FOR STUDY AND REFERENCE. — In addition to other works on physical culture mentioned in this book, the student will find the following books of much value and interest: 1. Lagrange's Physiology of Exercise. 2. Anderson's Light Gymnastics. 3. Blaikie's How to get Strong. 4. Enebuske's Gymnastic Days' Orders. 5. Mara Pratt's New Calisthenics. 6. Mara Pratt's Supplement to New Calisthenics. 7. Roberts's Gymnastic Exercises Classified. 8. Maclaren's Physical Education. 9. Ladies' Home Calisthenics.

Reference has been made to several standard works on the Swedish system on page 65.

# GLOSSARY

The ordinary use of a glossary is to explain, in some detail, many of the more difficult words used in the text. The pupil will, however, find it an admirable and profitable review exercise, to drill himself on the spelling, derivation, and definition of each word, mastering a few words every day.

---

Ab-do'men (Latin *abdo, abdere*, to conceal). The largest cavity of the body, containing the liver, stomach, intestines, etc.

Ab-sor'bents (L. *absorbere*, to suck up). The vessels which take part in the process of absorption.

Ab-sorp'tion. The process of sucking up nutritive or waste matters by the blood-vessels, or lymphatics.

Ac-com-mo-da'tion of the Eye. The alteration in the shape of the crystalline lens, which accommodates, or adjusts, the eye for near and remote vision.

Ac-e-tab'u-lum (L. *acetum*, vinegar). A little cup used by the ancients for holding vinegar ; applied, in anatomy, to the round cavity in which the hip-bone receives the head of the femur.

Ac'id (L. *acidus*, from *acere*, to be sour). A substance usually sour, sharp, or biting to the taste.

Ad'am's Ap'ple. An angular projection of cartilage in the front of the neck. It is particularly prominent in males, and is so called from a notion that it was caused by the apple sticking in the throat of our first parent.

Al-bu'men, or Albumin (L. *albus*, white). An animal substance resembling the white of an egg.

Al-bu'min-oids. A class of substances resembling albumen : they may be derived from either the animal or vegetable kingdoms.

Al'i-ment (L. *alo*, to nourish). That which affords nourishment ; food.

Al-i-ment'a-ry Ca-nal (from *aliment*). A long tube in which the food is digested, or prepared for reception into the blood.

Al'ka-li (Arabic, *al kali*, the soda plant). A name given to certain substances, such as soda, potash, and the like, which have the power of combining with acids.

391

An-æs-thet'ics (**Greek** *an*, without, *aisthesia*, feeling). Those medicinal agents which prevent the feeling of pain, such as chloroform, ether, laughing-gas, etc.

A-nat'o-my (Gr. *anatemno*, to cut up). The science which describes the structure of living things.

A-or'ta (Gr. *aeirein*, to lift up). The largest artery of the body, arising from the left ventricle of the heart. The name was first applied to the two large branches of the trachea, which appear to be lifted up by the heart.

A'que-ous Humor (L. *aqua*, water). The watery fluid occupying the space between the cornea and crystalline lens of the eye.

A-rach'noid Mem'brane (Gr. *arachne*, a spider, and *eidos*, like). The thin covering of the brain and spinal cord, between the dura mater and the pia mater.

Ar'bor Vi'tæ (L.). Literally, "the tree of life;" a name given to the peculiar appearance presented by a section of the cerebellum.

Ar'ter-y (Gr. *aer*, air, and *tereo*, to contain). A vessel by which blood is carried away from the heart. It was supposed by the ancients to contain only air; hence the name.

Ar-tic-u-la'tion (L. *articulo*, to form a joint). The more or less movable union of bones, etc.; a joint.

A-ryt'e-noid Car'ti-la-ges (Gr. *arutaina*, a ladle). Two small cartilages of the larynx, resembling the mouth of a pitcher.

As-sim-i-la'tion (L. *ad*, to, and *similis*, like). The conversion of food into living tissue.

Au'di-to-ry (L. *audito*, to hear) Nerve. One of the cranial nerves: it is the special nerve of hearing.

Au'ri-cle (L. *auricula*, a little ear). A cavity of the heart.

Bac-te'ria (Gr. *baktron*, a staff). A microscopic, vegetable organism; certain species are active agents in fermentation, while others appear to be the cause of infectious diseases.

Bile. The gall, or peculiar secretion of the liver; a viscid, yellowish fluid, and very bitter to the taste.

Blad'der (Saxon, *bleddra*, a bladder, a goblet). A bag, or sac, serving as receptacle of some secreted fluid; as the *gall-bladder*, etc. In common language, the receptacle of the urine in man and other animals.

Bright's Dis-ease'. A group of diseases of the kidney, first described by Dr. Bright.

Bronch'i (Gr. *bronchos*, windpipe). The two first divisions, or branches, of the trachea: one enters each lung.

Bronch'i-al Tubes. The smaller branches of the trachea within the substance of the lungs, terminating in the air-cells.

Bronch-i'tis (from *bronchos*, and *-itis*, a suffix signifying inflammation of the larger bronchial tubes; a "cold" affecting the lungs.

**Bun'ion.** An enlargement and inflammation at the first joint of the great toe.

**Cal'lus** (L. *calleo*, to be thick-skinned). Any excessive hardness of the skin, caused by friction or pressure.

**Ca-nal'** (L. *canalis*, a canal). In the body, any tube or passage.

**Ca'pil-la-ry** (L. *capillus*, hair). The smallest blood-vessels, so called because they are so tiny.

**Car-bon'ic A'cid.** The gas which is present in the air breathed out from the lungs; a waste product of the animal kingdom, and a food of the vegetable kingdom. More correctly called *Carbon Dioxide*.

**Car'di ac** (Gr. *kardia*, the heart). The cardiac orifice of the stomach is the upper one, and is near the heart; hence its name.

**Car-niv'o-rous** (L. *ca'ro*, flesh, and *vo'ro*, to devour). Subsisting upon flesh.

**Car'ron Oil.** A mixture of equal parts of linseed-oil and lime-water, so called because first used at the Carron Iron Works in Scotland.

**Car'ti-lage.** A tough but flexible material, forming a part of the joints, air-passages, nostrils, ear; gristle.

**Ca'se-ine** (L. *ca'seus*, cheese). The albuminoid substance of milk: it forms the basis of cheese.

**Ca-tarrh'.** An inflammation of a mucous membrane.

**Cell** (L. *cella*, a storeroom). The name of the tiny microscopic elements, which, with slender threads or fibres, make up most of the body: they were once believed to be little hollow chambers; hence the name.

**Cem'ent.** The substance which forms the outer part of the fang of a tooth.

**Cer-e-bel'lum** (diminutive for *cer'-ebrum*, the brain). The little brain, situated beneath the posterior third of the cerebrum.

**Cer'e-brum** (L.). The brain proper, occupying the entire upper portion of the skull. It is nearly divided into two equal parts, called "hemispheres," by a cleft extending from before backward.

**Chlo'ral.** A powerful drug used by physicians to induce sleep.

**Cho'roid** (Gr. *chorion*, skin, and *eidos*, form). The middle coat of the eyeball.

**Chyle** (Gr. *chulos*, juice). The milk-like fluid formed by the digestion of fatty articles of food in the intestines.

**Chyme** (Gr. *chumos*, juice). The pulpy liquid formed by digestion in the stomach.

**Cil'i-a** (pl. of *cil'i-um*, an eyelash). Minute hair-like processes found upon the cells of the air-passages and other parts that are commonly moist.

**Cir-cu-la'tion** (L. *cir'culus*, a circle). The course of the blood through the blood-vessels of the body.

Co-ag-u-la'tion (L. *coag'ulo*, to curdle). Applied to the process by which the blood clots or solidifies.

Coch'le-a (L. *coch'lea*, a snail-shell). The spiral cavity of the internal ear.

Conch'a (Gr. *konche*, a mussel-shell). The shell-shaped portion of the external ear.

Con-ges'tion (L. *con*, together, and *gero*, to bring). An unnatural gathering of blood in any part of the body.

Con-junc-ti'va (L. *con*, together, and *jun'go*, to join). A thin layer of mucous membrane which lines the eyelids, and covers the front of the eyeball, thus joining the latter to the lids.

Con-nect'ive Tis'sue. The network which connects the minute parts of most of the structures of the body.

Con-sti-pa'tion (L. *con*, together, and *stipo*, to crowd close). Costiveness; tardiness in evacuating the bowels.

Con-sump'tion (L. *consumo*, to consume.) A disease of the lungs, attended with a fever and cough, and causing a gradual decay of the bodily powers. The medical name is *phthisis*.

Con-trac-til'i-ty (L. *con*, together, and *tra'ho*, to draw). The property of a muscle which enables it to contract, or draw its extremities closer together.

Con-vo-lu'tions (L. *con*, together, and *vol'vo*, to roll). The tortuous foldings of the external surface of the brain.

Con-vul'sion (L. *convel'lo*, to pull together). A more or less violent agitation of the limbs or body.

Corn (L. *cor'nu*, a horn). A small portion of the outer skin, of horn-like hardness.

Cor'ne-a (L. *cor'nu*, a horn). The transparent, horn-like substance which covers a part of the front of the eyeball.

Cor'pus-cles Blood. (L. dim. of *cor'pus*, a body). The small disks which give to the blood its red color : the *white* corpuscles are globular and larger.

Cos-met'ic (Gr. *kosmeo*, to adorn). Applied to articles which are supposed to increase the beauty of the skin.

Cra'ni-al (L. *cra'nium*, the skull). Pertaining to the skull.

Cri'coid (Gr. *kri'kos*, a ring, and *eidos*, form). A cartilage of the larynx, resembling a seal-ring in shape.

Crys'tal-line Lens (L. *crystal'lum*, a crystal). One of the so-called humors of the eye; a double-convex body situated in the front part of the eyeball.

Cu'ti-cle (L. dim. of *cu'tis* the skin). The scarf-skin; also called the *epider'mis*.

Cu'tis (Gr. *skutos*, a skin, or hide). The true skin, lying beneath the cuticle; also called the *der'mis*.

Dan'druff. The small scales, or particles, which separate from the scarf-skin of the scalp.

De-cus-sa'tion (L. *decus'sis*, the Roman numeral ten, X.). A reciprocal crossing of fibres from side to side.

De-gen-er-a'tion (L. *degenerare*, to grow worse; to deteriorate). A change in the structure of any organ which makes it less fit to perform its duty or function.

Deg-lu-ti'tion (L. *deglutire*, to swallow down). The act, or process, of swallowing.

De-lir'i-um. A state in which the ideas of a person are wild, irregular, and unconnected.

Den'tine (L. *dens, dentis*, a tooth). The hard substance which forms most of a tooth; ivory.

De-o-do-ri'zer. An agent which corrects any foul or unwholesome odor.

Di'a-phragm (Gr. *diaphrasso*, to divide by a partition). A large, thin muscle which separates the cavity of the chest from the abdomen.

Di-ar-rhœ'a (Gr. *diarrhein*, to flow through). An unnaturally frequent and liquid evacuation of the bowels.

Dif-fus'ion of Gases. The power of gases to become intimately mingled.

Dis-in-fect'ant. Agents used to destroy the causes of infection.

Dis-lo-ca'tion (L. *dislocare*, to put out of place). The name of an injury to a joint, in which the bones are displaced or forced out of their sockets.

Dis-sec'tion (L. *dis*, apart, and *seco*, to cut). The cutting up of an animal in order to learn its structure.

Duct (L. *du'co*, to lead). A narrow tube : the *thoracic duct* is the main trunk of the absorbent vessels.

Du-o-de'num (L. *duode'ni*, twelve). The first division of the small intestines, about twelve fingers'-breadth long.

Dys-pep'si-a (Gr. *dus*, ill, *peptein*, to digest). A condition of the alimentary canal in which it digests imperfectly. Indigestion.

El'e-ment. One of the simplest parts of which anything consists.

E-met'ic (Gr. *emeo*, to vomit). Having power to excite vomiting. Also, a medicine which causes vomiting.

E-mul'sion (L *emulgere*, to milk). Oil in a finely divided state, suspended in water.

En-am'el (French, *émail*). The dense material which covers the crown of a tooth.

En'e-ma (L. *enema*, a clyster). An injection thrown into the rectum, as a medicine, or to impart nourishment.

Ep-i-glot'tis (Gr. *epi*, upon, and *glottis*, the entrance to the windpipe). A leaf-shaped piece of cartilage which covers the top of the larynx during the act of swallowing.

Ep'i-lep-sy (Gr. *epilepsis*, a seizure). A nervous disease accompanied by fits in which consciousness is lost. The falling sickness.

Eu-sta'chi-an (from an Italian anatomist named Eustachi). The tube which leads from the throat to the middle ear, or tympanum.

**Ex-cre'tion** (L. *excer'no*, to separate). The separation from the blood of the waste matters of the body; also, the materials excreted.

**Ex-pi-ra'tion** (L. *expi'ro*, to breathe out). The act of forcing air out of the lungs.

**Ex-ten'sion** (L. *ex*, out, and *ten'do*, to stretch). The act of restoring a limb, etc., to its natural position after it has been flexed, or bent; the opposite of *flexion*.

**Fau'ces.** The part of the mouth which opens into the pharynx.

**Fer'ment.** That which causes fermentation, as yeast.

**Fer-men-ta'tion** (L. *fermentum*, boiling, hot). The process of undergoing an effervescent change, as by the action of yeast; in a wider sense, the change of organized substances into new compounds by the action of a ferment. It differs in kind according to the nature of the ferment which causes it.

**Fe'nes-tra** (L.). Literally, a window; the opening between the middle and internal ear.

**Fi'bre** (L. *fibra*, a filament). One of the tiny threads of which many parts of the body are composed.

**Fi'brine** (L. *fi'bra*, a fibre). An albuminoid substance found in the blood.

**Flex'ion** (L *flecto*, to bend). The act of bending a limb, etc.

**Fol'li-cle** (L. dim. of *fol'lis*, a money-bag). A little pouch, or depression, in a membrane.

**Fo-men-ta'tion** (L. *fo'veo*, to keep warm). The application of any warm, soft medicinal subtance to some part of the body by which the vessels are relaxed.

**Fo-ra'men.** A hole, or aperture.

**Fu-mi-ga'tion** (L. *fu'migo*, to perfume a place). The use of certain fumes to counteract contagious effluvia.

**Func'tion** (L. *functio*, a doing). The special duty of any organ of the body.

**Fun'gous Growths** (L. *fun'gus*, a mushroom). A low grade of vegetable life.

**Gan'gli-on** (Gr. *ganglion*, a knot). A knot-like swelling in a nerve; a smaller nerve-centre.

**Gas'tric** (Gr. *gaster*, stomach). Pertaining to the stomach.

**Gel'a-tine** (L. *gelo*, to congeal). An animal substance which dissolves in hot water, and forms a jelly on cooling.

**Germ** (L. *germen*, a sprout, bud). Disease germ; a name applied to certain tiny bacterial organisms which have been demonstrated to be the cause of disease.

**Gland** (L. *glans*, an acorn). An organ consisting of follicles and ducts, with numerous blood-vessels interwoven.

**Glot'tis** (Gr. *glotta*, the tongue). The narrow opening between the vocal cords.

**Glu'ten.** Literally, glue; the glutinous albuminoid ingredient of wheat.

**Groin.** The lower part of the abdomen just above each thigh.

Gus-ta'tion (L. *gusto*, to taste). The sense of taste.

Gus'ta-to-ry Nerve. The nerve of taste supplying the front part of the tongue, a branch of the "fifth" pair.

Gym-nas'tics (Gr. *gumnazo*, to exercise). The practice of athletic exercises.

Hem'i-spheres (Gr. *hemi*, half, and *sphaira*, a sphere). Half a sphere, the lateral halves of the cerebrum, or brain proper.

Hem'or-rhage (Gr. *hai'ma*, blood, and *regnumi*, to burst). Bleeding, or the loss of blood.

He-pat'ic (Gr. *hepar*, the liver). Pertaining to the liver.

Her-biv-o'rous (L. *her'ba*, an herb, and *vo'ro*, to devour). Applied to animals that subsist upon vegetable food.

Hic'cough. A convulsive motion of some of the muscles used in breathing, accompanied by a shutting of the glottis.

Hu'mor. Moisture; the humors are transparent contents of the eyeball.

Hy-dro-pho'bi-a (Gr. *hudor*, water, and *phobeomai*, to fear). A disease caused by the bite of a rabid dog or other animal.

Hy'gi-ene (Gr. *hygieia*, health). The art of preserving health, and preventing disease.

In-ci'sor (L. *inci'do*, to cut). Applied to the four front teeth of both jaws, which have sharp, cutting edges.

In'cus. An anvil; the name of one of the bones of the middle ear.

In'di-an Hemp. The common name of *Cannabis Indica*, an intoxicating drug, known as *hasheesh*, and by many other names, in Eastern countries.

In-fe'ri-or Ve'na Ca'va (L.). Lower hollow vein; the chief vein of the lower part of the body.

In-flam-ma'tion (L. prefix *in*, and *flammo*, to flame). A redness or swelling of any part of the body, with heat and pain.

In-sal-i-va'tion (L. *in*, and *sali'va*, the fluid of the mouth). The mingling of the saliva with the food during the act of chewing.

In-spi-ra'tion (L. *inspi'ro*, *spira'-tum*, to breathe in). The act of drawing in the breath.

In-teg'u-ment (L. *inte'go*, to cover). The skin, or outer covering of the body.

In-tes'tine (L. *in'tus*, within). The part of the alimentary canal which is continuous with the lower end of the stomach; also called the bowels.

I'ris (L. *i'ris*, the rainbow). The thin, muscular ring which lies between the cornea and crystalline lens, and which gives the eye its special color.

Jaun'dice (Fr. *jaunisse*, yellow). A disorder in which the skin and eyes assume a yellowish color.

Lab'y-rinth. The internal ear, so named from its many windings.

Lach'ry-mal Ap-pa-ra'tus (L. *lach'ryma*, a tear). The organs for forming and carrying away the tears.

**Lac'te-als** (L. *lac, lac'tis,* milk). The absorbent vessels of the small intestines.

**Lar'ynx.** The cartilaginous tube situated at the top of the wind-pipe, or trachea; the organ of the voice.

**Lens.** Literally, a lentil; a piece of transparent glass or other substance so shaped as either to converge or disperse the rays of light.

**Lig'a-ment** (L. *li'go,* to bind). A strong, fibrous material binding bones or other solid parts together: it is especially necessary to give strength to joints.

**Lig'a-ture** (L. *ligo,* to bind). A thread of some material used in tying a cut or injured artery.

**Lobe.** A round, projecting part of an organ, as of the liver, lungs, or brain.

**Lymph** (L. *lym'pha,* pure water). The watery fluid conveyed by the lymphatic vessels.

**Lym-phat'ic Ves'sels.** A system of absorbent vessels.

**Mal'le-us.** Literally, the mallet; one of the small bones of the middle ear.

**Mar'row.** The soft, fatty substance contained in the central cavities of the bones.

**Mas-ti-ca-tion** (L. *mas'tico,* to chew). The act of cutting and grinding the food to pieces by means of the teeth.

**Me-a'tus** (L. *meare,* to wander). A passage or channel.

**Me-dul'la Ob-lon-ga'ta.** The "oblong marrow," or nervous cord, which is continuous with the spinal cord within the skull.

**Mem-bra'na Tym'pan-i.** Literally, the membrane of the drum; a delicate partition separating the outer from the middle ear; it is sometimes incorrectly called the drum of the ear.

**Mem'brane.** A thin layer of tissue serving to cover some part of the body.

**Mi'cro-scope** (Gr. *mikros,* small, and *skopeo,* to look at). An optical instrument which assists in the examination of minute objects.

**Mi'crobe** (Gr. *mikros,* little, and *bios,* life). A microscopic organism particularly applied to bacteria.

**Mo'lar** (L. *mo'la,* a mill). The name applied to the three back teeth at each side of the jaw; the grinders, or mill-like teeth.

**Mo'tor** (L. *mo'veo, mo'tum,* to move). Causing motion; the name of those nerves which conduct to the muscles the stimulus which causes them to contract.

**Mu'cous Mem-brane.** The thin layer of tissue which covers those internal cavities or passages which communicate with the external air.

**Mu'cus.** The glairy fluid which is secreted by mucous membranes, serving to keep them in a moist condition.

**My-o'pi-a** (Gr. *muo,* to shut, and *ops,* the eye). A defect of vision dependent upon an eyeball that is too long, rendering distant objects indistinct; near-sight.

Nar-cot'ic (Gr. *narkoo*, to benumb). A medicine, which, in poisonous doses, produces stupor, convulsions, and sometimes death.

Na'sal (L. *na'sus*, the nose). Pertaining to the nose.

Nerve Cell. A minute, round, and ashen-gray cell found in the brain and other nervous centres.

Nerve Fi'bre. An exceedingly slender thread of nervous tissue.

Nos'tril (Anglo-Saxon, *nosu*, nose, and *thyrl*, a hole). One of the two outer openings of the nose.

Nu-cle'o-lus (diminutive of *nu'-cleus*). A little nucleus.

Nu'cle-us (L. *nux*, a nut). A central part of any body, or that about which matter is collected. In anatomy, a cell within a cell.

Nu-tri'tion (L. *nu'trio*, to nourish). The processes by which the nourishment of the body is accomplished.

O-don'toid (Gr. *odous*, a tooth, *eidos*, shape). The name of the bony peg of the second vertebra, around which the first turns.

Œ-soph'a-gus. Literally, that which carries food. The tube leading from the throat to the stomach; the gullet.

O-le-ag'i-nous (L. *o'leum*, oil). Of the nature of oil; applied to an important group of food-principles, — the fats.

Ol-fac'to-ry (L. *olfa'cio*, to smell). Pertaining to the sense of smell.

Op'tic (Gr. *opto*, to see). Pertaining to the sense of sight.

Or'bit (L. *or'bis*, a circle). The bony socket or cavity in which the eyeball is situated.

Or'gan (L. *organum*, an instrument or implement). A portion of the body having some special function or duty.

Os'se-ous (L. *os*, a bone). Consisting of, or resembling, bone.

Pal'ate (L. *pala'tum*, the palate). The roof of the mouth, consisting of the hard and soft palate.

Pal-pi-ta'tion (L. *palpitatio*, a frequent or throbbing motion). A violent and irregular beating of the heart.

Pa-pil'la. The name of the small elevations found on the skin and mucous membranes.

Pa-ral'y-sis (Gr. *paraluein*, to loosen; also, to disable). Loss of function, especially of motion or feeling. Palsy.

Par'a-site. A plant or animal that grows or lives on another.

Pa-tel'la (L. dim. of *pat'ina*, a pan). The knee-pan.

Pel'vis. Literally, a basin. The bony cavity at the lower part of the trunk.

Pep'sin (Gr. *pepto*, to digest). The active principle of the gastric juice.

Per-i-car'di-um (Gr. *peri*, about, and *kardia*, heart). The sac enclosing the heart.

Per-i-os'te-um (Gr. *peri*, around, *osteon*, a bone). A fibrous membrane which surrounds the bones.

Per-i-stal'tic Move'ments (Gr. *peri*, round, and *stello*, to send). The slow, wave-like movements of the stomach and intestines.

**Per-i-to-ne'um** (Gr. *periteino*, to stretch around). The investing membrane of the stomach, intestines, and other abdominal organs.

**Per-spi-ra'tion** (L. *perspi'ro*, to breathe through). The sweat.

**Pe'trous** (Gr. *petra*, a rock). The name of the hard portion of the temporal bone, in which is situated the drum of the ear and labyrinth.

**Pha-lan'ges** (Gr. *phalanx*, a body of soldiers closely arranged in ranks and files). The bones of the fingers and toes.

**Phar'ynx** (Gr. *pharunx*, the throat). The cavity between the back of the mouth and gullet.

**Phys-i-ol'o-gy** (Gr. *phusis*, nature, and *logos*, a discourse). The science of the functions of living, organized beings.

**Pi'a Ma'ter** (L.). Literally, the tender mother; the innermost of the three coverings of the brain. It is thin and delicate; hence the name.

**Pig'ment** (L. *pingo*, to paint). Coloring-matter.

**Plas'ma** (Gr. *plasso*, to mould). Anything formed or moulded. The liquid part of the blood.

**Pleu'ra** (Gr. *pleura*, the side, also a rib). A membrane covering the lung, and lining the chest.

**Pleu'ri-sy.** An inflammation affecting the pleura.

**Pneu-mo-gas'tric** (Gr. *pneumon*, the lungs, and *gaster*, the stomach). It is the principal nerve of respiration; also called the *vagus*, or wandering nerve.

**Pneu-mo'nia.** An inflammation affecting the air-cells of the lungs.

**Poi'son** (Fr. *poison*). Any substance, which, when applied externally, or taken into the stomach or the blood, works such a change in the animal economy as to produce disease or death.

**Por'tal Vein** (L. *porta*, a gate). The venous trunk formed by the union of all the veins coming from the intestines. It carries the blood to the liver.

**Pres-by-o'pi-a** (Gr. *presbus*, old, and *ops*, the eye). A defect of the accommodation of the eye, caused by the hardening of the crystalline lens; the "far-sight" of adults and aged persons.

**Proc'ess** (L. *proce'do, proces'sus*, to proceed, to go forth). Any projection from a surface; also, a method of performance; a procedure.

**Pro'te-id** (Gr. *protos*, first, and *eidos*, form). An element allied to nitrogen; a substance containing such elements; an albuminoid.

**Pty'a-lin** (Gr. *ptualon*, saliva). The peculiar organic ingredient of the saliva.

**Pul'mo-na-ry** (L. *pul'mo, pulmo'nis*, the lungs). Pertaining to the lungs.

**Pulse** (L. *pel'lo, pul'sum*, to beat). The striking of an artery against the finger, occasioned by the contraction of the heart, commonly felt at the wrist.

**Pu'pil** (L. *pupil'la*). The central,

round opening in the iris, through which light passes into the interior of the eye.

**Py-lo′rus** (Gr. *puloros*, a gatekeeper). The lower opening of the stomach, at the beginning of the small intestine.

**Re′flex** (L. *reflexus*, turned back). The name given to involuntary movements produced by an excitation travelling along a sensory to a centre, where it is turned back or reflected along motor nerves.

**Re′nal** (L. *renes*, the kidneys). Pertaining to the kidneys.

**Res-pi-ra′tion** (L. *respi′ro*, to breathe frequently). The function of breathing, comprising two acts, — *inspiration*, or breathing in; and *expiration*, or breathing out.

**Ret′i-na** (L. *re′te*, a net). The innermost of the three tunics, or coats, of the eyeball, being an expansion of the optic nerve.

**Sac′cha-rine** (L. *sac′charum*, sugar). Of the nature of sugar; applied to the group of food-substances which embraces the different varieties of sugar, starch, and gum.

**Sa-li′va.** The moisture, or fluids, of the mouth, secreted by the salivary glands. The spittle.

**Scle-rot′ic** (Gr. *skleros*, hard). The tough, fibrous, outer coat of the eyeball.

**Se-ba′ceous** (L. *sebum*, fat). Resembling fat; the name of the oily secretion by which the skin is kept flexible and soft.

**Se-cre′tion** (L. *secer′no, secre′tum*, to separate). The process of separating from the blood some essential, important fluid; which fluid is also called a secretion.

**Sem-i-cir′cu-lar Ca-nals.** Three canals in the internal ear.

**Sem-i-lu′nar** (L. *semi*, half, *luna*, mooned). Shaped like a half-moon.

**Sen-sa′tion.** The perception of an external impression by the nervous system.

**Se′rum.** The clear, watery fluid which separates from the clot of the blood.

**Sock′et** (L. *soccus*, a kind of low-heeled shoe). An opening into which anything is fitted.

**Spasm** (Gr. *spasmos*, convulsion). A sudden, violent, and involuntary contraction of one or more muscles.

**Spe′cial Sense.** A sense by which we receive particular sensations, differing from those of general sensibility; such as those of sight, hearing, taste, and smell.

**Sprain.** An injury to the ligaments or tendons about a joint.

**Sta′pes.** Literally, a stirrup; one of the small bones of the middle ear, shaped somewhat like a stirrup.

**Stim′u-lant** (L. *stimulo*, to prick or goad on). An agent which causes an increase of vital activity in the body or any of its parts.

**Stri′a-ted** (L. *strio*, to furnish with channels). Marked with fine, parallel lines.

**Sub-cla′vi-an Vein** (L. *sub*, under,

and *clavis*, a key). The great vein bringing back the blood from the arm and side of the head; so called because it is situated underneath the *clavicle*, or collar-bone.

**Su-pe′ri-or Ve′na Ca′va** (L. upper hollow vein). The great vein of the upper part of the body.

**Sut′ure** (L. *sutura*, a seam). The union of certain bones of the skull by the interlocking of jagged edges.

**Sym-pa-thet′ic Sys′tem of Nerves.** A double chain of nervous ganglia, connected by numerous small nerves, situated chiefly in front of, and on each side of, the spinal column.

**Symp′tom** (Gr. *sun*, with, and *pipto*, to fall). A sign, or token, of disease.

**Syn-ov′i-al** (Gr. *sun*, with, *oön*, an egg). The liquid which lubricates the joints; joint-oil. So called from its resemblance to the white of a raw egg.

**Sys-tem′ic.** Belonging to the system, or body, as a whole.

**Sys′to-le** (Gr. *sustello*, to contract). The contraction of the heart by which the blood is expelled from that organ.

**Tac′tile** (L. *tac′tus*, touch). Relating to the sense of touch.

**Tar′tar.** A hard crust which forms on the teeth, and is composed of salivary mucous, animal matter, and a compound of lime.

**Tem′po-ral** (L. *tem′pus*, time, and *tem′pora*, the temples). Pertaining to the temples; so called be-cause the hair begins to turn white with age in that portion of the scalp.

**Ten′don** (L. *ten′do*, to stretch). The white, fibrous cord, or band, by which a muscle is attached to a bone; a sinew.

**Tet′a-nus** (Gr. *teino*, to stretch). A disease marked by persistent contractions of all or some of the voluntary muscles; those of the jaw are sometimes solely affected; the disorder is then termed locked-jaw.

**Tho′rax** (Gr. *thorax*, a breast-plate). The upper cavity of the trunk of the body, containing the lungs, heart, etc.; the chest.

**Thy′roid** (Gr. *thureos*, a shield, and *eidos*, form). The largest of the cartilages of the larynx: the projection in the front of the neck is called "Adam's apple."

**Tis′sue.** Any substance or texture in the body formed of various elements, such as cells, fibres, blood-vessels, etc., interwoven with each other.

**To-bac′co** (Indian, *tabaco*, the tube, or pipe, in which the Indians smoked the plant). A plant used for smoking and chewing, and in snuff. It has a strong smell and a pungent taste.

**Tra′che-a** (Gr. *trachus*, rough). The windpipe, or the largest of the air-passages.

**Trans-fu′sion** (L. *transfun′do*, to pour from one vessel to another). The operation of injecting blood taken from one person into the veins of another: other fluids than blood are sometimes used.

**Trich-i'na Spi-ra'lis.** (A twisted hair). A minute species of parasite, or worm, which infests the flesh of the hog: may be introduced into the human system by eating pork not thoroughly cooked.

**Tym'pa-num** (Gr. *tumpanon*, a drum). The cavity of the middle ear, resembling a drum in being closed by two membranes.

**U-re'a** (L. *urina*, urine). One of the principal constituents of the urine, secreted from the blood by the kidneys.

**U-re'ter** (Gr. *ourein*, to pass water). The excretory duct of the kidneys.

**U'vu-la** (L. *uva*, a grape). The small, pendulous body attached to the back part of the palate.

**Var'i-cose** (L. *varix*, a dilated vein). An unhealthily distended or enlarged vein.

**Vas'cu-lar** (L. *vasculum*, a little vessel). Pertaining to, or possessing blood or lymph vessels.

**Ve'nous** (L. *ve'na*, a vein). Pertaining to, or contained within, a vein.

**Ven-ti-la'tion.** The introduction of fresh air into a room or building in such a manner as to keep the air within it in a pure condition.

**Ven'tri-cles of the Heart.** The two largest cavities of the heart, situated at its apex, or point.

**Ver'te-bral Col'umn** (L. *ver'tebra*, a joint). The backbone ; also called the spinal column and spine.

**Ves'ti-bule.** A portion of the interior ear, communicating with the semicircular canals and the cochlea ; so called from its fancied resemblance to the vestibule, or porch, of a house.

**Vil'li** (L. *vil'lus*, shaggy hair). Minute, thread-like projections found upon the internal surface of the small intestine, giving it a velvety appearance.

**Vit're-ous** (L. *vi'trum*, glass). Having the appearance of glass; applied to the humor occupying the largest part of the cavity of the eyeball.

**Viv-i-sec'tion** (L. *vi'vus*, alive, and *se'co*, to cut). The practice of operating upon living animals, for the purpose of studying some physiological process.

**Vo'cal Cords.** Two elastic bands, or ridges, situated in the larynx; the essential parts of the organ of voice.

# INDEX

# ADVERTISEMENTS

# NATURAL SCIENCE TEXT-BOOKS.

**Principles of Physics.** A Text-book for High Schools and Academies. By ALFRED P. GAGE, *Instructor of Physics in the English High School, Boston.* $1.30.

**Elements of Physics.** A Text-book for High Schools and Academies. By ALFRED P. GAGE. $1.12.

**Introduction to Physical Science.** By ALFRED P. GAGE. $1.00.

**Physical Laboratory Manual and Note-Book.** By ALFRED P. GAGE. 35 cents.

**Introduction to Chemical Science.** By R. P. WILLIAMS, *Instructor in Chemistry in the English High School, Boston.* 80 cents.

**Laboratory Manual of General Chemistry.** By R. P. WILLIAMS. 25 cents.

**Chemical Experiments.** General and Analytical. By R. P. WILLIAMS. For the use of students in the laboratory. 50 cents.

**Elementary Chemistry.** By GEORGE R. WHITE, *Instructor of Chemistry, Phillips Exeter Academy.* $1.00.

**General Astronomy.** A Text-book for Colleges and Technical Schools. By CHARLES A. YOUNG, *Professor of Astronomy in the College of New Jersey,* and author of "The Sun," etc. $2.25.

**Elements of Astronomy.** A Text-book for High Schools and Academies, with a Uranography. By Professor CHARLES A. YOUNG. $1.40. **Uranography.** 30 cents.

**Lessons in Astronomy.** Including Uranography. By Professor CHARLES A. YOUNG. Prepared for schools that desire a brief course free from mathematics. $1.20.

**An Introduction to Spherical and Practical Astronomy.** By DASCOM GREENE, *Professor of Mathematics and Astronomy in the Rensselaer Polytechnic Institute, Troy, N.Y.* $1.50.

**Elements of Structural and Systematic Botany.** For High Schools and Elementary College Courses. By DOUGLAS HOUGHTON CAMPBELL, *Professor of Botany in the Leland Stanford Junior University.* $1.12.

**Elements of Botany.** By J. Y. BERGEN, Jr., *Instructor in Biology in the English High School, Boston.* $1.10.

**Laboratory Course in Physical Measurements.** By W. C. SABINE, *Instructor in Harvard University.* $1.25.

**Elementary Meteorology.** By WILLIAM M. DAVIS, *Professor of Physical Geography in Harvard University.* With maps, charts, and exercises. $2.50.

**Blaisdell's Physiologies:** Our Bodies and How We Live, 65 cents; How to Keep Well, 45 cents; Child's Book of Health, 30 cents.

**A Hygienic Physiology.** For the Use of Schools. By D. F. LINCOLN, M.D., author of "School and Industrial Hygiene," etc. 80 cents.

*Copies will be sent, postpaid, to teachers for examination on receipt of the introduction prices given above.*

**GINN & COMPANY, Publishers, Boston, New York, Chicago, Atlanta.**

# THE BEST HISTORIES.

**Myers's History of Greece.** — Introduction price, $1.25.

**Myers's Eastern Nations and Greece.** — Introduction price, $1.00.

**Allen's Short History of the Roman People.** — Introduction price, $1.00.

**Myers and Allen's Ancient History.** — Introduction price, $1.50. This book consists of Myers's Eastern Nations and Greece and Allen's History of Rome bound together.

**Myers's History of Rome.** — Introduction price, $1.00.

**Myers's Ancient History.** — Introduction price, $1.50. This book consists of Myers's Eastern Nations and Greece and Myers's History of Rome bound together.

**Myers's Mediæval and Modern History.** — Introduction price, $1.50.

**Myers's General History.** — Introduction price, $1.50.

**Emerton's Introduction to the Study of the Middle Ages.** — Introduction price, $1.12.

**Emerton's Mediæval Europe (814-1300).** — Introduction price, $1.50.

**Fielden's Short Constitutional History of England.** — Introduction price, $1.25.

**Montgomery's Leading Facts of English History.** — Introduction price, $1.12.

**Montgomery's Leading Facts of French History.** — Introduction price, $1.12.

**Montgomery's Beginner's American History.** — Introduction price, 60 cents.

**Montgomery's Leading Facts of American History.** — Introduction price, $1.00.

**Cooper, Estill and Lemmon's History of Our Country.** — Introduction price, $1.00.

For the most part, these books are furnished with colored and sketch maps, illustrations, tables, summaries, analyses and other helps for teachers and students.

## GINN & COMPANY, Publishers.

BOSTON.  NEW YORK.  CHICAGO.  ATLANTA.